LORD KELVIN
The Dynamic Victorian

LORD KELVIN

The Dynamic Victorian

Harold Issadore Sharlin
in collaboration with Tiby Sharlin

The Pennsylvania State University Press

University Park and London

Library of Congress Cataloging in Publication Data

Sharlin, Harold I.
Lord Kelvin, the dynamic Victorian.

Includes bibliographical references and index.
1. Kelvin, William Thomson, Baron, 1824–1907.
2. Physicists—Great Britain—Biography.   3. Science—
Great Britain—History.   I.   Sharlin, Tiby.   II.   Title.
QC16.K3S53   530'.092'4 [B]   78-50771
ISBN 0-271-00203-4

We wish to thank Miss Margaret Tait,
the Smithsonian Institution, and the Library
of Congress for permission to use the photographs
that appear in this book.

To the memory of
Jennie Sharlin
and
Rose Mintz

CONTENTS

PREFACE

An impetus behind this book was the belief that biographies of scientists can have an important integrative and explanatory function in the history of science. This goal is elusive, for in writing scientists' biographies, it is necessary to explain the technical achievements of the scientists in question, and few historians feel themselves equal to this task. Thus biographies of scientists are usually written by other scientists who become ensnared in the problem of explaining the science and neglect the subject's development and significance. The difficulties of presenting technical information are exaggerated while the difficulties of writing biography are underrated. In the field of biography of scientists, the situation is similar to that of the history of science as a whole. A historian of science should be, some insist, a historian and a scientist. Since few people are equally competent in two fields, the mixture is often that of taking a scientist and adding a dash of history. Facts about the past, after they are interpreted, become history. The interpretation makes the past intelligible and, above all, relevant to the living. In *Dynamic Victorian,* the biography of a great scientist who was also an engineer, I have tried to integrate the subject's life and work and to place the result in a broader historical context that would illuminate the history of science and technology in the nineteenth century.

William Thomson, later Lord Kelvin, was the foremost figure of the age that saw the quest of classical physics concluded and marked the beginning of the modern era of atomic physics, quantum mechanics, and relativity. His contributions began with the analogy between heat and electricity. This analogy formed part of the foundation for concentrating on energy as the basis for modern physics. Thomson directly influenced James Clerk Maxwell, whose work culminated in the electromagnetic theory of light, the theory which ushered in the modern period of electrical theory and electrical technology. Thomson made several other fundamental contributions to the abstract mathematical science of the nineteenth century, primarily in the fields of electricity and thermodynamics.

What makes Thomson's life so useful in understanding scientific and technological culture is the way his activities extended in all directions beyond the development of scientific theory. His letters, diaries, and mathematical notebooks show how published and written communications between scientists modified and extended their theoretical insights. All the major figures in nineteenth-century physics—British, Continental, and American—came within Thomson's purview.

His relations with his father, his family, and the educational experiences in a Scottish and an English university were the ingredients of Thomson's individualistic approach to the scientific questions of the day. Science is more than the abstract interpretation of nature; it is the product of personalities living in an age that fosters particular attitudes. Victorian science bore the mark of the age, and that association partly explains the rejection of some of those theories by the new age of twentieth-century science.

William Thomson's most important contribution as a technological innovator was to the successful completion of the Atlantic cable. Since he was an engineering interpreter of his own scientific theories, the way in which abstract mathematical theory is converted into concrete technological devices forms part of this biography. His interpretation of the second law of thermodynamics brought him into conflict with the evolutionary geologists and biologists of the day. That controversy illuminates the impact of scientific theory on religious and popular thought.

Thomson fits into no simple definition of scientist-engineer. The reader of this biography will be introduced to an extraordinary figure of a past era who in no way fits the image of the modern specialist. It is just this characteristic of Thomson's life that will take readers, including scientists, engineers, and those who know little science, into the wider universe of technological innovation derived from scientific theory. His work on the Atlantic cable shortened the time period between Europe and America from weeks to seconds. His controversy with the Darwinians concerning the age of the earth put him "on the side of the angels" in one of the few scientific debates that the Victorian public followed. His discovery of a universal tendency in nature toward the dissipation of energy provided an idea that Henry Adams and others translated into predictions of doom for the earth—a prediction that is once again current. William Thomson was the nonpareil scientist of the nineteenth century, and the amount of original material available made it possible to write a biography that encompasses the most important scientific changes of the nineteenth century.

Preface

How does an imaginative, intuitive, as well as individualistic insight become part of the accepted scientific credo? Beginning with Thomson's mathematical analogy between heat and electricity (devised when he was seventeen), I have tried to show that there was a logical progression to the dynamical theory—a dominant theme in late nineteenth-century physics. A number of other scientists, including Faraday, Maxwell, Joule, and Stokes, were instrumental in making the dynamical view acceptable. But, as Maxwell pointed out, the mathematical analogy between heat and electricity was an unlikely analogy and something only Thomson could have thought of.

Thomson's extraordinary ability to comprehend both the experimental and mathematical view of a subject in a sort of stereoscopic vision cast him in the role of synthesizer at critical junctures in nineteenth-century physics. As the physicist's physicist and the moving spirit whose unique imagination brought together the disparate theories of thermodynamics and electricity, Thomson's position in the nineteenth-century revolution in physics can be compared with Newton's role in the seventeenth century and Einstein's in the twentieth.

One way to see a shift in scientific thinking is by picturing the new school of thought that is vanquishing the old. In this type of melodrama, Kelvin serves as the foil to twentieth-century physics. But his letters and papers show him to be a serious critic who had to be dealt with before the new school of scientists could be assured of the correctness of their theories. Their decision, which was the only way to avoid Kelvin's challenge, was to forgo the nineteenth-century hope for a unified theory of science. As later events demonstrated, Kelvin's and his contemporaries' hope was suspended, not lost. Kelvin's dream, in modern relativistic form, was revived by Einstein in his later years, and he too was regarded as an obstructionist.

Because of the immediacy with which scientists view their history, they dismiss the work of most nineteenth-century scientists as being outmoded or quaint. James Clerk Maxwell is the glaring exception: twentieth-century scientists look upon his equations as "beautiful." But he lived only until 1879—not long enough to oppose the changes that occured in physics at the turn of the century. Lord Kelvin lived into the twentieth century—long enough to be blamed for holding up progress. As is the case with leading figures, his reputation has been the basis for a number of myths: in his own time his prowess was exaggerated, and in more recent times his reputation has been denigrated by various critics.

In writing this biography, it has not been my intent to cover all of nineteenth-century physics, nor have I dealt in detail with all of Thomson's papers. An analysis of his large corpus of work can be of interest only to physicists and to the limited extent that there may be suggestions in Thomson's research that will be of use to people working on the same problem. From the point of view of what is important in Thomson's work to present-day research, one selects a different list of papers than Thomson's contemporaries would. For example, the Thomson effect and vortex motion are important in some scientific circles today, although in the nineteenth century these theories could not have measured large. I have selected for analysis those papers which were important in Thomson's intellectual development and in nineteenth-century science. Therefore, I make no apologies to physicists who pick up this biography expecting to see an analysis of what they *now* designate "important" papers. Since this work is directed toward a historically interested audience (be they historians, scientists, or educated laymen), I express Thomson's ideas in the language of words and not that of mathematics. Although the language of mathematics might help certain scientists understand Thomson's work, the biography would be limited to technically trained readers.

Mathematics is clearly the medium for scientific creativity. I do not believe, however, that anything is lost in transposing the medium to the printed word for purposes of historical explanation. If I have succeeded, the reader will know more about the origins of nineteenth-century science, why certain theories were satisfactory explanations in their time, and, by implication, why scientists today find them unsatisfactory.

Silvanus P. Thompson's *Life of William Thomson, Baron Kelvin of Largs* has the eulogistic character of a book in which the biographer is part of the same milieu as his subject. As Thompson wrote, "Not in our generation will it be possible to exercise dispassionate vision or to disentangle the ultimate from the obvious."[1] It is an irony that the Thompson biography has provided the material for a number of false or misleading stories about William Thomson. Kelvin (as he was known after 1892) has been used by scholars as an example of nineteenth-century obstructionism to modern physics. He was, as the evidence in my work shows, rather among those who were most impressed by the discovery of X rays and radioactivity. His attempt to fit these discoveries into his dynamical scheme failed, but he did not doubt the authenticity of the laboratory research.

I have attempted to use this biography of a man of so many talents as an integrative device. Biography joins together what specialization has torn asunder. In telling the story of a scientist's life, one moves naturally from the things that influenced him (influence of a culture on science), to the way he formulated his ideas in conjunction with others (the history of scientific theory), to the interpretation of these ideas by other scientists, technologists, and popularizers (the diffusion of scientific thought). In Thomson's case, since he lived to the end of a scientific era, his life was part of a significant shift in fundamental concepts (a scientific revolution).

Although several books have dealt with Thomson's life, none has approached S. P. Thompson's biography of him in the amount of detailed information. It was a constant reference throughout the writing of this book. Thompson provided a wealth of material including a complete bibliography of Thomson's published works and a list of his patents. I cannot say that I am writing this biography because I found a cache of letters or diaries not known to S. P. Thompson, although I did see, through the courtesy of several people, a number of letters that I am sure he did not. But the justification for a new biography must be that the subject needs to be treated from a different viewpoint.

In 1965, I received a grant from the Iowa State University Research Foundation and two supplementary grants from the American Philosophical Society enabling me to do a year of research in Great Britain and making possible the photocopying of the greater part of Kelvin's papers. I gladly acknowledge my debt to those who were responsible for making the awards. The bulk of Kelvin's papers is deposited in the Cambridge University Library, whose staff made it possible for Tiby Sharlin and me to accomplish a great deal in the course of the year's work. I thank H. R. Creswick, the librarian, for permission to make use of the original material. Several people made our stay in Cambridge a particularly happy and profitable one. I single out for thanks Dr. Mary B. Hesse, Dr. Michael Hoskin, and Mr. Gerd Buchdahl. The staff of the Whipple Science Museum was unusually helpful in both personal and professional matters, and I thank them.

Mr. J. T. Lloyd of the University of Glasgow was helpful to me in my search and made my visits to Glasgow pleasant. Dr. R. T. Hutcheson, the registrar at the University of Glasgow, was generous with his time. Personnel at the Royal Society, the Institution of Electrical Engineers, the British Museum, and the National Register of Archives gave me much appreciated aid. Mr. Allan Jeffreys of the University of Keele supplied me with information about manuscript sources in Britain.

Preface

I also wish to thank Mr. and Mrs. James K. Bottomley for welcoming me in their home and for giving me access to their collection of Thomson's papers. Other descendants of Thomson's, Mr. J. S. Vesey Brown of Bermuda and Mr. P. Graham Blandy of Madeira, supplied me with information and papers. Miss Margaret Tait, P. G. Tait's granddaughter, generously provided me with many letters in the Tait-Thomson correspondence, with other correspondence, as well as pages from a notebook that Tait kept. She has also given me some photographs. I cannot thank her enough. Mrs. L. Ursell helped me translate some of the German letters written to Thomson, for which I am very grateful.

Iowa State University awarded me several summer grants to continue the writing of this book. Most of the early drafts were written in the university library—a good place to work and think. In addition, an exceptionally resourceful library staff (special thanks to Miss Elizabeth A. Windsor, Mrs. Mildred E. McHone, and Miss Eleanor J. McKee) were no end of help to me. The Niels Bohr Library of the Center for History of Physics at the American Institute of Physics supplied me with useful information, as did Mr. Samuel Suratt, archivist of the Smithsonian Institution. Professor Thomas S. Kuhn generously made arrangements for me to have an office at Princeton University, where I was able to work for another summer on the book.

Professor John C. Greene first suggested that I do a biography of William Thomson. Professor Stephen G. Brush read an early draft of the manuscript. Dr. Hesse again put me in her debt when she read chapter 6. Her many suggestions were most helpful. Professor Robert A. Leacock helped me to see in words the ideas expressed in Thomson's mathematical analogy between heat and electricity. I am grateful to Dr. Ivor Grattan-Guinness for his critical reading of the manuscript.

Our children, Allan, Joshua, and Shifra, have endured and participated in the making and debating of this book. Allan N. Sharlin has given me valuable professional advice.

A very stimulating year as Fellow at the Woodrow Wilson International Center for Scholars in Washington, D.C., gave me an opportunity to take another look at the manuscript. I benefited from many conversations there. Mr. David Wise gave helpful advice. Dr. Jules Gueron generously volunteered to read an earlier draft of the manuscript and made many helpful suggestions.

Chapters 6 and 10 were published in a different form in *Annals of Science*.

CHAPTER ONE

Paternal Momentum

The father-son relationship has an element that is analogous to the historical basis of scientific research. The son has his father to contend with, and he rejects him at his peril. The scientific tradition may obstruct modern science, but to deny that tradition entirely is to undermine the basis for scientific investigation. For the son and a new generation of scientists, there are two courses open: submit to the past and be a duplicate hemmed in by the lessons of someone else's experience, or escape. Those who seek to escape the past without doing violence to the historical relationship between the present and the past are able to maintain their independence and make original contributions.

James Thomson was his son's teacher, his son's mentor, and the source from which his son derived his perseverance. At a critical point in his adolescence, William Thomson diverged from the path set for him by his father and avoided becoming a duplicate of the man who shaped his intellect and personality. An ambitious father can direct his son's career along certain established paths of endeavor, but the course that William Thomson followed was beyond his father's range of vision and beyond his rule. James Thomson was a self-made man; the momentum he gathered in bettering himself became the power behind his son's success.

Annaghmore, the family farm near Belfast, had been worked by the Thomson family for almost two hundred years. James Thomson was the fifth child in a family of six. He was taught to read by his sisters, but he taught himself mathematics. He was fascinated by the sundial in front of his house and, at the age of eleven, tried to make one. When it did not keep correct time, he patiently began to work out the principle of dialing, which involved geometric astronomy. He had difficulty in spacing the dial's hour markings and had to calculate the unevenly spaced distances between marks. The first step was to work out the spaces for Annaghmore's latitude. He thor-

1

oughly mastered the mathematical principles involved and was able to make vertically and horizontally placed dials. Eventually, he made dials set at any angle.[1]

James Thomson had no one to guide him through the difficulties of his early mathematical studies, and no one with whom to exchange ideas, but he did have a sympathetic older brother who was interested. When he asked how sundials worked, James Thomson succeeded in teaching him the rudiments of dialing. That kind of reinforcement was all that James Thomson needed; he began to harbor thoughts of life beyond the confines of the farm. He was allowed to go to a nearby school run by a Presbyterian minister, where he studied classics and mathematics. In 1810, at the advanced age of twenty-four, James Thomson went to Glasgow and entered the university. In 1904, his son William recounted his father's arduous journey to Glasgow:

> There were no steamers, nor railways, nor motor cars in those days. Can young persons of the present time imagine life to be possible under such conditions? My father and his comrade-students, chiefly aspirants for the ministry of the Presbyterian Synod of Ulster and for the medical profession in the North of Ireland, had to cross the Channel twice a year in whatever sailing craft they could find to take them. *Once* my father was fortunate enough to get a passage in a Revenue cutter, which took him from Belfast to Greenock in ten hours. Another of his crossings was in an old smack whose regular duty was to carry lime, not students, from Ireland to Scotland. The passage took three or four days. . . . [2]

The Scottish university system had neither entrance requirements nor entrance examinations, and offered an educational opportunity to those with little preparation who, like Thomson, sought to improve their station in life. There were many avenues open in the expanding Scottish industrial economy. In this respect, the Scottish universities served as a model for the colleges of agriculture and mechanic arts (land-grant colleges) in the United States. The same Scottish universities provided an opportunity for young men thoroughly grounded in their studies, such as James's son William, to advance to the highest level of research and to the most important scientific positions in Britain. It was a university system matched to its society.[3]

For a four-year period, James Thomson divided his time; from May to November he taught mathematics and classics at the local boys' school, and from November to May he attended the University of Glasgow. He finished the course in theology and completed most of the medical curriculum. James took his degree in 1814, an unusual step in that day because a degree from Glasgow was low in public esteem, that is, it was common knowledge that the B.A. as well as the M.A. were awarded without examination to anyone who paid the fees or at most to candidates who took an examination that was easily passed. The degree, however, made Thomson eligible for a parish position in Ireland. But when he was offered the position of arithmetic and geography teacher at the new Royal Belfast Academical Institution, he readily accepted. He had finally found a livelihood for which he was well suited. When the institution expanded by adding a college, Thomson was appointed to the mathematical chair, which he held simultaneously with the mastership.[4]

Before he met his first mathematics class of the day at 8:00 A.M., Thomson, his daughter reported, had been up for four hours, with a pot of coffee at his side, writing one of his textbooks. The books were an exceptionally successful undertaking. His first book, *A Treatise on Arithmetic in Theory and Practice*, published in 1819, went into seventy-two editions. His *Introduction to Modern Geography* and *Romance of the Heavens* were both published in 1827. Thomson's success was apparent. In 1829, he received an honorary doctorate from the University of Glasgow. There must have been a great deal of celebration in the Thomson household at the news of this honor; it was one of William Thomson's earliest memories. In 1830, James Thomson published *Elements of Plane and Spherical Trigonometry*. A fourth edition was published in 1844. In 1831, he finished *Introduction to the Differential and Integral Calculus*, an indication of the advanced level of his mathematics. A second edition of the book appeared in 1848.[5] He also edited an edition of Euclid's *Elements of Geometry*, and he wrote *Algebra* in 1834. Like his other books, *Algebra* was an exemplar of clarity and introduced new methods.

In 1817, James Thomson married Margaret Gardner, the daughter of a Glasgow merchant. He was thirty-one, and his bride was twenty-seven. They had seven children, three girls and four boys. Elizabeth, the eldest, was born in 1818. William, the fourth child and the second oldest boy, was born on 26 June 1824. One girl did not survive childhood, so the Thomson family consisted of two girls, who were the oldest, and four boys.

Margaret Gardner Thomson became ill soon after the birth of her seventh child, Robert, in February 1829. Unwilling to be parted from her family, she did not take the recuperative trip prescribed by her physician. Her health remained precarious, and in January 1830 she was confined to her bed. In May she died. Elizabeth tells how the children were told of their mother's death:

> The five elder children were taken down to our father's study. He was sitting there alone, at the side of the fire. As the little troop came into the room, he opened his arms wide, and we ran into them, and he clasped us all to his heart. . . . And he said with a choking voice "You have no Mamma now." He held us a long time so; his whole breast heaving with convulsive sobs. Then he gathered the two little ones, William and John, on his knees, and kept his arms tight round us all,—his head resting on the cluster of young heads closely pressed together; and there we remained in silence and darkness, except for the glow of the dying embers. . . . [6]

William was six years old when his mother died.

James Thomson did not remarry. Agnes Gall, Margaret Thomson's sister, was separated from her husband and was staying with the Thomsons at the time of her elder sister's death. Aunt Agnes became housekeeper to the family. Apparently, the arrangement was a happy one.

In addition to her household duties, it was the mother's responsibility to teach the girls. As was common, most of the children's education was provided at home while they received some supplementary education at school. Distinctions in the home education were made according to expectations. A division was made along male-female lines, as it was in one of the children's favorite books, *Harry and Lucy*, by Maria Edgeworth.[7] Being only girls, Elizabeth and Anna were not exposed to the more difficult subjects which their brothers studied. Dr. Thomson taught the two eldest boys, James and William; Anna and Elizabeth taught the youngest, John and Robert. How much could they learn from their sisters? How much was expected of them? The system probably benefited the girls more than their students. Taking her task seriously, Elizabeth read Locke on education in order to find out how to make her lessons more appealing to Robert.[8]

This was the nineteenth-century form of primogeniture. The eldest males were taught the most useful subjects and learned from the best teacher, their father. The outcome was predictable. In another Victorian family, James Mill taught his eldest son, John Stuart Mill. At the age of three, Mill began to study Greek, and at the age of eight, he began to study Latin. When he was little more than a child himself, he was required to teach the younger children in the family. According to his father, Mill was a prodigy not because of any unusual innate intelligence, but because of his teaching.[9]

4

Dr. Thomson was a more indulgent father than James Mill. In fact, he was unusual in the conscientious way he cared for his children. Fate burdened him with two handicaps: his wife's death and the recurring illnesses of all his children, except William. Elizabeth contracted scarlet fever when she was five, suffered severe nosebleeds when she was fifteen, and was confined to bed one summer with migraine headaches.[10] When shifted to his children, his own ambitions were frustrated by the retarding effect of their illnesses. Robert was truly ill fated. He was subject to fits soon after birth, had two operations for calculus (stones), and contracted scarlet fever. His studies were often interrupted by fevers, and at least one of them was near fatal. John, born in 1826, was bedridden with rheumatism at the age of seventeen.

James failed to complete his first two apprenticeships in engineering because of illness. The first time, he was made lame from hiking in poorly fitted shoes. During his second apprenticeship, he had a chronic cold. But when he transferred to a firm in Manchester, his cold went away. Nevertheless, he was forced to return home because of a serious heart ailment. The symptoms were a very rapid pulse, 120 beats per minute, and a general feeling of weakness. These were anxious months for Dr. Thomson. He was told that his eldest son, twenty-three years of age, would die from heart disease in a few months. The diagnosis proved wrong, but James was ill for months after that.[11]

Dr. Thomson's concern for his three sons' health replaced any ambitions he may have had for them. How could he hope for anything for James except that he not be an invalid for the rest of his life? William might have received his father's attention by default, but he was not passive in the father-son relationship.

In addition to his good health, William was an unusually handsome child. Elizabeth, his chief admirer, said she enjoyed looking at him in his crib because he was so pretty. When William was two, Elizabeth recalled, an artist asked to use him as a model for an angel. At the age of seven, he was to receive a first prize at school, and Elizabeth insisted on washing and dressing him for the occasion. In spite of his indignance, she managed to overcome his opposition and dressed him in white trousers and a black surtout. How could William help being a little vain? One day Elizabeth found him admiring himself in a looking glass and saying to himself, "P'itty b'ue eyes Willie Thomson got!"[12]

In the children's competition for their father's affection and approval, William was the most successful. Elizabeth thought William was a "great pet" of their father's. She believed Dr. Thomson favored William "partly, perhaps, on account of his extreme beauty, partly on account of his wonderful quickness of apprehension, but most of all, I think, on account of his coaxing, fascinating ways, and the caresses he lavished on his 'darling papa.'" She also remembered that not "any of us were ever in the slightest degree jealous of William on account of our father's making him a little more of a pet than the rest of us. We were proud of him, and indeed we thought the child petted the father even more than the father petted the child, but we saw plainly that the fondling of his little son pleased him." It was William's habit from infancy to fling himself onto his father when he returned home and to kiss and caress him fondly. William's show of affection took the place of words when he was an infant, but he continued the habit of baby talk even after he spoke very well. The other children thought this device was an affectation and laughed at William, but he ignored them.[13]

In the winter of 1832, Dr. Thomson was appointed professor of mathematics at Glasgow University. He delayed moving the family to Glasgow because of the cholera epidemic that was there. But they moved the following October and settled in a house in Professors' Court.[14] Dr. Thomson must have wondered if he made the right move when he learned that he would be earning far less than he anticipated because his salary was dependent on the size of his class. During his first year at Glasgow he earned "less than nothing," but the next year his daughter reported:

He announced an afternoon course of lectures for ladies, on geography and astronomy, to be given twice a week in his class-room. Such a thing had never been heard of before in the University, and it was extremely popular. The large class-room was crowded in every corner, and it was a novel and interesting spectacle to see bench rising above bench filled with fashionably-dressed ladies, everyone looking intent, and many taking notes. All the belles of Glasgow were among the students. This class was carried on for two or three years with undiminished popularity, till the pressure of other engagements compelled my father to give it up. . . . [15]

A kind of sibling competitiveness carried over to school work when James and William began classes at the University of Glasgow in 1832. William was eight years old at the time and James ten. Both brothers formally matriculated in 1834. Ordinarily, the earliest age for matriculation was fourteen, but James and William were far better prepared than the majority of students entering the university regardless of age. Each class had different criteria for winning prizes, and awards were decided by class vote. William won a prize in 1836 for translating Lucian's *Dialogues of the Gods* with a full parsing of the first three dialogues. The brothers stood at the head of the list of the monthly voluntary examinations given by their father in his mathematics class. In the competition for class prizes, they easily surpassed the other boys, and William always surpassed his brother. William won first prize in both the junior and senior mathematical classes; James won second prize both times. In the 1838 natural philosophy class, William won first prize, and James again was second.[16]

Did James resent William? There is no indication that he did. Nor is there any reason to believe that he harbored such feelings. Perhaps the lack of resentment was due to Dr. Thomson's attempts to bolster his eldest son's ego by such acts as giving him "a pair of scales as a reward for taking the specific gravity of things, since he did not get a prize."[17] However, the consequence of James's university experience was that he was unable to establish his independence by excelling in his studies. That route was cut off by William. James chose engineering as a field far enough from William's and his father's interests.

Where James did attempt to establish his independence was in religion. After his father died, he became a Unitarian. William felt no need for such rebellion, and if he were to rebel it would not be through religion, which for William was never a matter for serious contemplation.

Elizabeth Thomson's description of six-year-old William's behavior at a Fundamentalist service forecast his balanced attitude toward religion. She had taken the younger children to Sunday services one morning when there happened to be a revival service. The congregation joined in with exclamations and groans and even threw Bibles in the air. William thought the scene was so funny that he had to smother his laughter and started the other boys laughing too. The minister noticed the lack of solemnity in the front pew and rebuked the boys who burst out laughing. Then the minister pointed to William and shouted, "Ye'll no lach when ye're in hell!" That was too much, and William rolled over onto the floor. Luckily, Dr. Thomson was not with the family that day as the red-faced Elizabeth hurried her group out of church.[18]

Like his brothers and sisters, William Thomson was content to accept passively his father's directions in all matters. Because William was Dr. Thomson's favorite, he tended to bind himself closer to his father. The tutelage might have been total had it not been for the latent conflict inherent in William's narcissism. He resembled the rest of the children in finding his father's dominance comfortable, but the acceptance was more evident in the girls' behavior. At the age of twenty-four, Elizabeth was reluctant to consider marriage since it meant leaving her father and family. She begged her suitor, Dr. David King, minister of Greyfriars' Secession (United Presbyterian) Church in Glasgow, not to mention marriage because she was "quite afraid of" that topic. "From my former letter," Elizabeth wrote, "you will perceive that I am in great doubt—indeed I feel both afraid and sad when I contemplate the possibility of any change in my lot. At home I am so very happy." When Elizabeth showed her father the letter she was going to send, Dr. Thomson asked to post the letter and enclosed a note of his own encouraging Dr. King, who eventually became Elizabeth's husband.[19]

Dr. Thomson's children were more cosmopolitan than he had been. He made sure of that. Everything he did impressed his children and seemed to contain a lesson. Elizabeth recalled a carriage ride when she was a girl of eight. It was a very hot day, and she remembered that the temperature was 80° only because her father had taken out his pocket thermometer and showed it to her. He explained that mercury expanded with heat and that the amount of heat was measured in degrees on the ivory scale. On the 1839 trip, William was not quite fifteen when he, his father, two sisters, three brothers, and a maid boarded a steamer to Liverpool and from there a train to London. Did William or his father set the following problem, which William carefully marked down in the diary he kept of the trip? "We start from Liverpool at half past 9. Another start from Birmingham two hours later. When should we meet?, the whole distance taking 5 hours. Ans. 1 oclock."[20]

In May 1839, Dr. Thomson and his family stopped in London prior to a trip to the Continent. They spent a month there waiting for Robert to recover from surgery for calculus. Robert suffered from severe attacks of pain which mystified the Glasgow doctors. They advised Dr. Thomson to consult with Sir Astley Cooper in London, who later informed Dr. Thomson that Robert would have to have an operation. Elizabeth was with Robert during his ordeal and reported that she saw "the poor little boy stretched, all bound, on the table before the operation began, and [she] stood on the mat outside the door while it was going on, listening to his moans. . . . There was no chloroform in those days to drown pain," she explained. During Robert's recovery, the family took in the gamut of sightseeing from the British Museum and the Tower of London to

Madame Tussaud's Waxworks; they saw the opera and theater as well.[21]

On William's second day at the British Museum, he saw the Elgin Marbles, observing that they were "so mutilated that they do not present great interest to visitors in general." But, as he noted in his diary, they should "be very interesting as studies to Artists." On another day, he went with his brothers James and John to the Adelaide Gallery of Practical Science and "was very much pleased with it. [He] saw a splendid exhibition of polarisation of light, and the oxy hydrogen microscope." There were also "a great many beautiful models of different pieces of machinery, among which were the electro-magnetic telegraph, the hydraulic telegraph, and the finest model of a marine engine [William] had ever seen."[22]

The family went for walks among the splendid luxury of London, and, as if they had not seen enough squalor in the streets of Glasgow, their father took them to the poorest sections of London to see, Elizabeth recalled, how the "ill-doing" lived. On the Continent, the boys were left in Paris to be tutored in French, and Dr. Thomson took the two girls on a tour of Switzerland. This was his way of compensating the girls for the lack of freedom they would experience later in life. Later on, the boys would have all the freedom they wanted for travel. On their return to Paris, the girls were fashionably outfitted before going back to Glasgow.[23]

All of the family, including Dr. Thomson, studied German the next year, and they spent the summer of 1840 in Germany. While spending a few days in London before embarking on their trip to Germany, William, James, and a friend went to the Polytechnic Institution, which was similar to the Adelaide Gallery of Practical Science, but newer. William thought the Polytechnic Institution was superior because it had "more of later inventions." The next morning, the family boarded a steamer for Rotterdam. They had "a rough passage, and a horrible tossing for nearly three hours" while waiting for the tide to come in. William recalled remaining in his "berth during all the time we lay off the bar, as I would infallibly have been sick if I had got up. The vessel rolled greatly from side to side, but the rolling was intermittent, as every two or three minutes it calmed down and then rose again, with perfect regularity. This probably arose from two sets of waves, of slightly different lengths coming in the same direction from different sources." They spent two days in Holland, then left for Germany. The boys were again expected to be more diligent in their study of the language. On this trip, the girls had dresses made from the best Genoese velvet. On the way home, they stopped in London to arrange James's apprenticeship at an engineering firm.[24]

For William Thomson, mathematics was the determined route to independence, but it was not certain he could surpass his father. When he entered the University of Glasgow, William began to study the same subjects and even had some of the same professors his father had in 1810. There was a possibility that William might become a carbon copy of his father, but no matter how diligent the father may be, or to what extent he controls his son's education, he cannot fabricate his son's ambitions for him. John Stuart Mill, whose experiences as a young man were similar to William Thomson's, thought "the sons of eminent fathers, who have spared no pains in their education, so often grow up mere parroters of what they have learnt, incapable of using their minds except in the furrows traced for them."[25] Paternal momentum has a way of continuing a son along the line taken by his father. William Thomson's boyhood presaged for him the same kind of life as his father's. He had the advantage of an earlier start and did not have to overcome the handicap of being self-taught as his father had been. William could continue in the same furrow, perhaps even going a greater distance because of his head start.

Chance now intervened in deciding William Thomson's future. In 1839, William Meikleham, the professor of natural philosophy who had held the chair since 1803, fell ill and his place was taken by John Pringle Nichol, the professor of astronomy. In his lectures, Meikleham emphasized Newton's mechanics and had his students read the difficult and by then standard works of Laplace and Lagrange. He had not altered the teaching of natural philosophy since the student days of Dr. James Thomson. When Nichol substituted in class, he introduced Fourier's new mathematical treatment of heat and the latest mathematical study of the theory of couples. Thomson, whose notebooks show that he was much more interested in natural philosophy than the other subjects he was required to take, was impressed by Nichol's discussion of the use of mathematics in understanding physical subjects.[26]

Professor Nichol introduced William to Joseph Fourier's *Théorie analytique de la chaleur*.[27] The book was a turning point in Thomson's life and had an influence on all of his subsequent scientific work. Unlike the older classics of science, such as Newton and Laplace, Fourier explained and justified his use of physical reasoning and mathematical intuition in dealing with the phenomena of heat. But even Fourier's book could not teach mathematical intuition. It is as impossible to teach mathematical intuition as it is to teach artistic creativity. Dr. Thomson was not an original mathematical thinker. As a teacher, he was an explicator or popularizer of mathematical ideas, and he could bring William only so far in his development. The acquiring of mathematical intuition is a psychic phenomenon, and in William Thomson's life it coincided with the time in which he had a pressing need to have some area free of his father's dominance.

Henri Poincaré said that mathematicians were of two kinds: logicians and intuitionists. It was by logic that a mathematician proved a proposition. Anyone can be taught to solve mathematical problems, but it is by intuition that the mathematician discovers new relations. Without intuition, Poincaré believed, the mathematician was like a writer who knew grammar but was destitute of ideas. The mathematician needed this "special aesthetic sensibility," or intuition, before he could do original work. Poincaré was no more successful than others in describing this mental attribute or how to acquire it. How does a mathematician know which combinations of all that he knows will be useful to him in a given setting? "The useful combinations are precisely the most beautiful," Poincaré wrote. "I mean those that can most charm that special sensibility that all mathematicians know, but of which laymen are so ignorant that they are often tempted to smile at it."[28]

William Thomson recalled that period of his life when he discovered Fourier as "a white era, an era of brightness in [his] memory."[29] In his autobiography, John Stuart Mill recalled that the first time he read a book by Bentham marked "an epoch in [his] life; one of the turning points in [his] mental history." Bentham's principle of utility was the keystone that held together the detached fragments of Mill's knowledge. "I now had opinions; a creed, a doctrine, a philosophy; in one among the best senses of the word, a religion; the inculcation and diffusion of which could be made the principal outward purpose of a life."[30]

Joseph Fourier was to William Thomson what Jeremy Bentham was to John Stuart Mill. A comparison shows how parallel the relations were. Both Mill and Thomson were about the same age when they read the book that greatly influenced their lives. Their delight transcended an appreciation of new ideas. Both of the young men had been encouraged and coached in their studies by their fathers. With the discovery of an additional parent, that is, an intellectual parent, Mill and Thomson became independent of all men. At that moment in their lives, they started a course of original work that diverged from the furrows of both their natural and intellectual parents. In each case the young man, without being conscious of what had happened, synthesized two different outlooks to manufacture a unique form that was thoroughly suited to his own unshaped thinking, just as his thinking was assuming a fixed, mature form.

Meikleham and Dr. Thomson had taught William how to *solve* problems. Nichol, through Fourier, opened the vista of original thought and mathematical intuition. Like all scientists, Thomson had to learn a body of knowedge. His father taught him well—he introduced him to the Leibnizian infinitesimal calculus that was still new in Great Britain. However, reading Fourier produced a transformation that opened the way, at a very early age, for William to become an original scientific thinker. Since this transformation had his father as a starting point, William Thomson's work in science was to be an integral part of his identity.

In the analogy between the historical basis for scientific research and the James Thomson-William Thomson father-son relationship, the comparison was much broader than the usual. For William, his father was the scientific tradition, and by transcending him William Thomson at one and the same time became independent of his father and was now prepared to make original contributions to scientific thought.

CHAPTER TWO

Originality

In the 1839–1840 school year, William Thomson won the astronomy class prize and was awarded a university medal for his "Essay on the Figure of the Earth." The professor of astronomy was John Pringle Nichol. In 1836, at the age of thirty-two, Nichol was appointed regius professor of astronomy at the University of Glasgow. In the years following his appointment, he became a prolific and successful author of popularizations of science as well as a popular lecturer. Nichol "appeared to take as much delight in explaining the laws that regulate the heavenly bodies to the unpretending mechanic as to the carefully-educated student." During 1848–1849, he gave a series of lectures in the United States.[1] Elizabeth describes another aspect of Nichol:

> The summer of 1838 was spent at Gourock, and from there Anna and I paid a delightful visit to Dr. and Mrs. Nichol, who had rented the Manse at the entrance of Glensannox in Arran. Dr. Nichol . . . used to take us frequently out in a boat, and we spent hours rowing about the rocky shore or lazily resting on our oars in some creek, while he read to us Tennyson's Poems, then recently published. It was a privilege to hear him, he brought out the meaning with such clearness and beauty by his sympathy and insight, and by the musical cadences of his voice.[2]

Speaking of his teacher many years later, Lord Kelvin said:

> The benefit we had from coming under his inspiring influence, that creative influence, that creative imagination, that power which makes structures of splendour and beauty out of the material of bare dry knowledge, cannot be overestimated.[3]

The title page of "An Essay on the Figure of the Earth" carries a quotation from Pope's *Essay on Man:*

> Mount where science guides;
> Go, *measure earth*, weigh air, and state the tides;

13

Instruct the planets in what orbs to run,
Correct old time, and regulate the sun.[4]

Thomson's preface to his essay listed the works of mathematicians that he had consulted: Airy, Poisson, Pontécoulant, Pratt, and Laplace. He readily admitted his debt to their works, but added that he had "throughout endeavoured to exercise his own judgement." He claimed originality in some of the modes of investigation and confessed that he had been anticipated in several instances. Thomson closed his preface with the piquant admission that he feared he may have been anticipated in other places as well.[5] The problem that the fifteen-year-old boy had chosen placed him in the company of a historical array of scientists. That he chose to contribute new ideas and seek new solutions instead of commenting on past work was a measure of his competence and his confidence.

That William Thomson had reached maturity in his mathematical thinking is revealed in his essay. At a number of points in it, his mathematics ran into an impasse: he found a "transcendental" equation insoluble by any known direct means, an equation that had never received a complete solution, and, at another point, he needed to know the resultant figure in order to solve the equation. Of course, the resultant figure of the earth was the objective. In such cases, his mature mathematical reasoning and the repertory of mathematical strategies used to escape the snares reveal that he possessed a singular mathematical intellect. He may not have been a Laplace as yet, but he had the potential for being front rank.

The problem of the shape or the figure of the earth was a "classical" one in science. The first to undertake this question from a mathematical view was Isaac Newton. He was followed by an illustrious array of mathematicians who refined or made alternative attempts at a most accurate estimate of the amount by which the earth differs from an oblate spheroid.[6] It seems that every scientist of mark, from Newton to the Astronomer Royal, had pondered the question. Why had no one found an answer? Because no final, exact answer is possible.

The introductory remarks in Thomson's "Essay on the Figure of the Earth" briefly recount the history of attempts to measure the shape of the earth. In these attempts, four different methods were used, and Thomson planned to develop each.[7] The first section, "Physical Theory," was the longest and most difficult. According to Newton's law of gravity, a fluid, such as the earth was when it was formed, naturally assumes a spherical shape. But the earth rotates, and the action of centrifugal force tends to flatten it at the poles. Just how much the earth was flattened at the poles depended on the relative magnitude of the two forces, gravity and centrifugal, and their resultant force.

Thomson used over fifty-six pages of his hand-written manuscript to discuss the figure of the earth by considering it as a fluid revolving around a fixed axis. The second section, "Disturbance in the Moon's Motion," applied Newton's universal law of gravitation to the attraction between the moon and the earth. The earth, were it a perfect sphere, would attract the moon uniformly at all points in its orbit around the earth, and that orbit would be a symmetrical geometric figure. The perturbations in the moon's path were due to the nonsphericity of the earth.[8]

In the third section of the essay, "Geodetic Measures," Thomson used the available measurements of short lengths of longitude and latitude to determine mathematically the overall dimensions of the earth. The fourth and shortest section, "Pendulum Observations," concluded the eighty-five-page essay. The swinging of a pendulum in different parts of the earth measures the variation in the gravity of the earth; these variations were still another way of approximating the nonsphericity of the earth.[9] Thomson's calculations agreed with values determined in other parts of the essay.

At the age of fifteen, Thomson achieved a great deal, as measured by the award of the university medal. The essay, however, meant more to him than that. He not only saved the manuscript, but he opened it periodically to rework problems, improving on the solution by new strategies. The dates of these emendations are associated with periods of stress in his life: 16 December 1844, when William wrote to his Aunt Agnes about his anxiety over the approaching Tripos; 13 September 1866, when he was on board the *Great Eastern*, returning to Glasgow after the successful laying of the Atlantic cable; and 21 October 1907, when Lady Kelvin was slowly recovering from a stroke.[10] A lasting imprint had been made on Thomson at the early age of fifteen. He remained attached to certain scientific questions and used the same type of method in solving the many different kinds of scientific questions that he encountered during his life.

In 1840, Dr. Nichol was the substitute natural philosophy professor. He talked to the class about the new ideas on heat contained in Joseph Fourier's book *Théorie analytique de la chaleur*. William asked Dr. Nichol if he thought a student could understand the book. Nichol admitted that he himself had not fully understood Fourier, but perhaps William might gain something by reading it. The subtle challenge was taken up, and William later recalled with delight, "In a fortnight I had mastered it—gone right through it." He had begun reading on the very day he received his university medal for "An Essay on the Figure of the Earth."[11]

The synthesis of the physical and the mathematical by Thomson came shortly after his introduction to two types of mathematical analysis: the application of Newton's law of attraction and Fourier's study of heat. In both cases, mathematics was used to determine the constant laws of nature. Experimentation, representing the physical interpretation of nature, was another investigatory technique in science. Up to the middle of the nineteenth century in the history of science, the laws discovered by one method of investigation were not expressed in terms easily understood by those using a different method.[12]

Fourier's mathematics was a central element of William Thomson's thought. To understand Thomson's subsequent research in science, it is necessary to know the significance he found in Fourier. And in order to do that, one must think in mathematical symbolism. However, to express Fourier through his equations alone is too austere intellectually, and to look at him in that way is to lose the manifold projections of his thought. This situation presents something of a dilemma in determining the significance of Fourier to Thomson, since Thomson saw and interpreted Fourier completely through mathematical symbols. But mathematical symbols do not adequately express the complexity of a man's thought, for symbols are only a representation, the bare bones as it were, and in spite of the limitations of expressing mathematical thought in words, it is only through words that we approach the wholeness of a thought. Fourier's work has been called a mathematical poem; mindful that the subtle meanings of a poem may be destroyed under analysis, the significance of Fourier to Thomson has to be expressed in words so that the reader can see the links between the thinking of the two men. For those who do not want an intervention into Thomson's thought, his collected mathematical papers may be consulted for a strictly mathematical presentation of his ideas.[13]

In the preface to his book, Fourier states that "primary causes are unknown to us; but are subject to simple and constant laws." This meant that he was avoiding hypotheses about the nature of heat, that is, the ultimate cause. Whether heat was a substance, caloric, or due to the motion of molecules, Fourier's analysis still held true. The mathematical relations he sought were between secondary effects of heat. He defined these secondary properties of bodies, which deter-

mined the action of heat, as the property of bodies to contain heat, to receive or transmit heat, or to conduct heat. Although each of these properties depended on the physical characteristics of the body and were affected by the nature of heat, the properties themselves were not physical properties. They were measurable characteristics which could be used to describe the motion of heat. The mathematical relations between these three basic properties described the successive thermal states of any solid body that might be selected. Fourier selected variations in terrestrial temperatures as being "one of the most beautiful applications of the theory of heat." The earth's absorption of solar heat varies with the time of day and the season of the year. In spite of these variations, underground temperatures of the earth remain constant, and Fourier sought a law to deduce these fixed underground temperatures from the variable temperatures observed at the surface.[14] The solution of this problem and the method used became the very basis upon which William Thomson later built his case on his estimation of the age of the earth.

The mathematics of Fourier, infinitesimal calculus, was similar to that used by Thomson in his "Essay on the Figure of the Earth." But where the shape of the earth was a question of matter acted upon by force, Fourier's work focused on force or energy itself. It was a mathematical method of conceptualizing energy and was the forerunner of the shifting emphasis to the characteristics and effects of energy that were considered separately from matter. When Thomson turned his attention to the questions connected with the behavior of an electric current in very long cables and, in particular, the Atlantic cable, he immediately saw this as an application of Fourier. Again, when systematizing the method for calculating the height of ocean tides acted upon by a complex of gravitational forces, he used Fourier's mathematics. The same mathematics was useful in many areas of his scientific research, hydrodynamics for example, and his second career in engineering was based to a great extent on the method of analysis he learned from Fourier. Fourier was to William Thomson what the telescope is to astronomers—a means of perceiving the physical world.

Fourier's method of analysis is remarkably general because it relies on expressing *arbitrary* conditions which are measurable, that is, conditions which are quantifiable and therefore mathematically describable. Being in no way dependent on the underlying physical cause, the mathematically determined properties may not be the true cause of the phenomena; indeed, Fourier denied that we knew they were, but the properties were a way of analyzing the differences in the movement of heat through solid bodies, *without* knowing the true physical cause. The motion of heat through all varieties of solid bodies can be compared by mathematical means. Any physical effect can be studied by the technique of abstraction; the success of the technique hinges on selecting what Fourier called "elementary properties."[15]

Fourier's book was first of all a mathematical treatise consisting of abstract reasoning. His postulates were of an idealized world, which were developed by following the line of mathematical logic. Fourier's work has become part of the basic lexicon of mathematics and has found followers who extend its ideas along strictly mathematical lines.[16] This course of thought is involuted and tends to separate itself from physical questions. Mathematics is one of the two pillars of modern science, and it is the physical applications of Fourier's mathematics that are of interest here.

To some scientists, mathematics is a tool enabling them to explore the physical world, and to others, mathematics is synonymous with the physical world. Fourier belonged to the second group: "Mathematical analysis has therefore necessary relations with sensible phenomena; its object is not created by human intelligence; it is a pre-existent element of the universal order, and is not in any way contingent or fortuitous; it is imprinted throughout all nature." Regardless of the view held, it is undeniable that mathematics is an incredibly successful conceptual scheme—an alternative to observation. Observation and experiment are in direct relation to the physical world and results obtained this way are the final test of ideas about the physical world. What Fourier did was to describe mathematically how heat acted. He claimed that mathematics had outstripped observation: "It has supplemented our senses, and has made us in a manner witnesses of regular and harmonic vibrations in the interior of bodies."[17]

Fourier's theory used mathematical equations for giving the instant-by-instant readings of imaginary thermometers distributed over a body through which heat passed.[18] He was describing a process, the passage of heat, that occurred in microcosm and that could not be observed by experiment. Mathematics did more than quantify heat; it described the *character* of the force by supposing how it acted. In the case of the effect of heat, the only possible way to link cause with effect, given the limitations of human observation, was through mathematical imagination. The uses found for Fourier's mathematical analysis revealed the possibilities of representing force.

Originality

The gradual unfolding of these possibilities was one of the most surprising developments of nineteenth-century science that extended well into the twentieth century. First heat, then, as a result of Thomson's work, electricity, magnetism, and light. The manifold occurrences in nature of harmonic motion, that is, any motion or force that varies according to a specific regular pattern, can be submitted to Fourier's mathematics. Fourier analysis is as lively a source for investigation today as it was in 1840. Thomson was one of the earliest British proponents of Fourier's ideas, even though he did not discover Fourier until eighteen years after the publication of *Théorie analytique de la chaleur.*

When the Thomson family left for Germany in May 1840, William took Fourier's book with him despite his father's injunction that

> all work should be left behind, so that the whole of our time should be given to learning German. We went to Frankfort [sic], where my father took a house for two months. The Nichols had lodgings adjacent, and came in to meals with us nearly every day. Now, just two days before leaving Glasgow I had got Kelland's book (*Theory of Heat,* 1837), and was shocked to be told that Fourier was mostly wrong. So I put Fourier into my box, and used in Frankfort [sic] to go down to the cellar surreptitiously every day to read a bit of Fourier. When my father discovered it he was not very severe upon me.[19]

Fourier's analysis had been labeled "nearly all erroneous" in a new book titled *Treatise on Heat,* by Philip Kelland, professor of mathematics at Edinburgh. After reading this book, William Thomson wrote his first published paper, "On Fourier's Expansions of Functions in Trigonometrical Series." The paper, which attacked Kelland and defended Fourier, was submitted anonymously by Dr. Thomson to the *Cambridge Mathematical Journal.* It appeared in the May 1841 issue and was signed with the pseudonym P. Q. R. In the article, the author stated that the material had been fully treated by Fourier, but since Fourier had not given a direct demonstration of his fundamental idea, many of the formulas were now believed to be in error.[20]

The first draft of the paper exhibited youthful directness. Thomson wanted it known that the paper "has been written, not for any wish to detract from the merit of Mr. Kelland's excellent treatise on heat, but for the purpose of doing justice to the celebrated philosopher to whom we are indebted for the mathematics and of removing any want of confidence in his work which might have been occasioned by Mr. Kelland's statement." In four-and-a-half hand-written pages, sixteen-year-old William refuted Kelland's criticism of Fourier. After giving several examples that verified Fourier, Thomson concluded: "I have examined the other series given by Fourier on this subject, and they seem to be all correct, with the exception of misprints and mistakes in transcription. . . . "[21]

D. F. Gregory, editor of the *Cambridge Mathematical Journal,* wrote to Dr. Thomson that in justice to Kelland "he should know the name of his opponent." A copy of the article was sent to Kelland, who accepted William's conclusions that Fourier's formulas were correct since Dr. Thomson was satisfied. Kelland admitted that he should have written in his book that Fourier's analysis was based on principles that were "not satisfactory." Now he only wanted the wording of William's remarks softened somewhat. A revised draft of the article was submitted to Kelland who, in a better mood, closed by telling Dr. Thomson that his son was on the right path and "will not fail to do great service to science." Although Dr. Thomson thought it was unnecessary for Kelland to see the final draft of the article, he asked the editor, Gregory, to send it to him anyway.[22] Rarely, it can be assumed, had a paper for the *Cambridge Mathematical Journal* been subjected to such delicate and complicated negotiations. Through it all, William allowed his father to manage the diplomatic details. It is unfortunate that at this juncture William Thomson did not learn to handle a controversy with firmness and finesse. It was a talent he never acquired.

Thomson's second article, "Note on a Passage in Fourier's Heat," answered another objection by Kelland.[23] Fourier had used a new mathematical relation (he expanded a function of x between arbitrary limits using an infinite sine series) to solve one of his problems in the movement of heat, but he had not demonstrated or proved that this relation was correct. Poisson found the same relation but used another method than Fourier had and objected to Fourier's solution. Thomson agreed with Kelland that Poisson had not surmounted the difficulty either, that is, Poisson's method did not give a satisfactory demonstration of the truth of Fourier's mathematical relation.[24] Just as Fourier could not demonstrate the correctness of his statement, so Kelland had no other argument than to say that sort of thing just was not done. Fourier's equations were not testable mathematically; the validity of his ideas could be demonstrated only by how well the equation described physical reality. The quantitative results obtained from Fourier's equations were verified by experimental data. That Fourier's insight was correct was proven by the physical results. Intuition had uncovered a physical truth. Was the mathematics correct, or was this a case of having arrived at the right answer using the wrong method? There was no doubt in Thomson's mind. His thinking fell right in with Fourier's intuition and, having accepted Fourier's postulates, he provided the proof that Fourier was unable to give: in three pages he demonstrated the validity of Fourier's mathematical inference and supplied the missing link in the chain of reasoning.

These two articles by Thomson, written in opposition to Kelland's doubts, were a common occurrence in the advance of mathematics. Fourier had an intuitive belief that his equations were possible, and he used the equations without demonstrating that they were mathematically correct. When mathematicians arrive at a statement intuitively, they bypass the ordinary steps of mathematical reasoning and are often unable to offer proof of their statements except to say that they "feel" they are right. The paradox of mathematics is that although it is founded on rigidly linked steps of logical analysis, it advances most often by using directly perceived truths independent of any reasoning process. Thomson's intuitive sense had not as yet ripened, and he could not at this point be called an original mathematician. He was, however, but one step away. By falling in so naturally with the intuitive thinking of Fourier, he was developing his own intuition. That Thomson was able to supply the proof, which Fourier and Poisson failed to do, indicated that he was able to follow the intuitive leap. By supplying the demonstration that was wanting, Thomson was playing the role of interpreter to Fourier.

It is significant that in 1840 so little attention was paid in Britain to Continental mathematics—almost twenty years after the publication of Fourier's treatise, virtually the only British writing on the new mathematics was Kelland's misinterpretation. On the Continent, the situation was much different. In 1812, the French Academy of Science proposed as its prize question: "To give the mathematical theory of the propagation of heat, and to compare this theory with exact observations." Fourier won the award, to no one's surprise, since it was known that he had already developed the necessary mathematics. In a paper given in 1807, Fourier had described his method of dealing with heat mathematically. Although his *Analytical Theory of Heat* was not published until 1822, his new ideas were appreciated before that. His election to the Académie des Sciences in 1817 was in recognition of his research on heat.[25] The French mathematicians' ready appreciation of Fourier contrasts with the British scientists' almost complete unawareness of the work being done in this area. In view of the importance of Fourier's mathematics to many areas of study, it might be thought that Thomson was born in the wrong country for the encouragement of his talents. But Britain had other advantages, as will be seen, that shaped Thomson's future.

In 1841, while at the family's summer house in Lermlash, Thomson wrote a third paper—the third segment of his core of originality; the two papers on Fourier and the "Essay on the Figure of the Earth" were the other two segments of that core. The paper marks the beginning of his original work and was a noticeable step beyond his other two papers and the essay. It was also provisional as so many first steps into originality are. When Dr. Thomson submitted the paper to the *Cambridge Mathematical Journal* on behalf of his son, who was still the anonymous P. Q. R. to the *Journal*'s readers, he apologized to Gregory for the paper's appearance; his son had not had the time to rewrite it.[26] Had William realized the implications of his new theory for nineteenth-century science, perhaps he would have found the time.

William's paper titled "On the Uniform Motion of Heat in Homogeneous Solid Bodies, and its Connection with the Mathematical Theory of Electricity" was published in February 1842.[27] The first part is a further development of Fourier's mathematical treatment of the flow of heat in solid bodies. The second part is on the theory of electrical attraction, which is a mathematical development, by analogy, of Newton's theory of gravitational attraction. William claimed that Fourier's equations for the movement of heat through a body could be used for solving problems in electrical attraction. The transformation of the equations from one use to the other was simply a mathematical operation.

An indication of the lasting quality of the paper was that it made a permanent imprint on advancing theory even though some of the ideas had been anticipated by other scientists. About a week after his father submitted the paper to Gregory, Thomson discovered that he had been anticipated. The list of people who had previously worked out some of the same results consisted of the most important contributors to scientific theory. The first was Michel Chasles, French mathematician, who had published two articles on the subject in French journals, which Thomson had not seen. Chasles used geometrical proofs almost exclusively, while Thomson used the infinitesimal calculus.[28] Later he found that Karl F. Gauss, a German mathematician, had also anticipated some of the ideas in the article. Only much later did Thomson find that George Green, a little-known English mathematician, had anticipated all three of them.

The number of anticipations is an indication of the complexity of Thomson's theory, since it touched on so many different ideas. The possibilities of Thomson's original insight were only glimpsed by him at the time. In 1854, when the article was reprinted, he realized somewhat better that the idea was a very fruitful one for research. Even then he did not recognize its full implications.[29]

Thomson's analogy between heat and electricity was a creative insight that did not occur to anyone else—including Chasles, Gauss, and Green who anticipated Thomson on other matters. James Clerk Maxwell, who more than any other scientist of the nineteenth century recognized the possibilities of Thomson's ideas, also recognized the extent of the analogy's originality. Maxwell wrote years later that the conduction of heat and the attraction of electricity were the most unlikely things to be treated analogously:

> In the one we have an attraction acting instantaneously at a distance, in the other heat creeping along from hotter to colder parts. The methods of investigation were also different. In the one the force on a given particle of electricity has to be determined as the resultant of the attraction of all the other particles. In the other we have to solve a certain partial differential equation which expresses a relation between the rates of variation of temperature in passing along lines drawn in three different directions through a point.[30]

The analogy was useful in a very fundamental way; it was, as Maxwell wrote, one of the most valuable of "science-forming ideas."[31] Once described, an analogy serves as a stimulus to thought, but that does not explain how Thomson happened to think of it. It is important to understand better his thought behind the theory because the same mode of thinking characterized all of his future research. As Maxwell recognized, the analogy was neither a physical one nor a strictly mathematical one. It was an abstract (nonphysical) comparison between things which were physical in their effects. In the real world, heat and electricity were physically dissimilar, yet at a certain level of abstraction away from the physical reality, they appeared to be analogous. By combining what was unique to Fourier with the kind of thinking employed by an experimentalist, Thomson had created a unique intellectual tool—the mathematico-physical analogy.

The paper "On the Uniform Motion of Heat," with its analogy between heat and electricity, was one of those diverging points in the progress of scientific theory whose scope cannot be appreciated until theory has moved some distance away. Thomson himself only began to recognize the significance of the idea three years later. At that time, he was told that the beliefs he held in electrical theory were in

23

logical opposition to one another; the mathematical theory of electricity could not be right if Faraday's experimental views were correct. Pressed to explain the anomaly of Faraday's results, Thomson declared that neither Faraday nor the mathematicians were wrong and, in fact, Faraday's experiments confirmed the mathematical theory. The difference in view arose from there being two methods of stating and interpreting physically the same laws. As a demonstration that there were two distinct ways of viewing the theory of electricity, Thomson cited the principles laid down in this paper.[32]

By the age of seventeen, William Thomson was an original thinker. For one so young and for the transformation to have occurred over a matter of months, it was strange that there was no outward sign of a drastic change having taken place. There was no outward sign because there was no inner struggle; Thomson had synthesized two aspects of his father's thinking, the physical and the mathematical, without sensing how sharply he had diverged from his father's intellectual path. To William Thomson, as to all original thinkers, it all seemed so obvious, although to an observer what he had accomplished was not less than a creative act.

In Dr. Thomson's teaching and writing, the physical and the mathematical views of nature were distinct but compatible. His textbooks on algebra, geometry, and the calculus represented his mathematical view; his books and popular lectures on geography and astronomy represented his physical view. Being a teacher but neither an experimentalist nor a theorist, Dr. Thomson introduced his son to these two views without biasing him toward either of these approaches to scientific work.

Dr. Thomson's mathematical imagination stopped short of his son's creative view. What was the source of William Thomson's mathematico-physical imagination? One source was John Pringle Nichol. In 1903, Lord Kelvin said of him: "In his lectures the creative imagination of the poet impressed youthful minds in a way that no amount of learning, no amount of mathematical skill alone, no amount of knowledge in science, could possibly have produced."[33] A more subtle source is found in an analogy between William Thomson's scientific beliefs and his father's religious views. Dr. Thomson's religion consisted of a set of abstract principles or morals that had a close relation to reality and to human behavior. He was a political and social reformer who was tireless in his efforts to eliminate the unjust requirement which forced faculty members to sign a religious oath. The requirement was counter to his concepts of Christianity. These religious concepts were applied by him in his authoritarian rule of his classes and his children. Although William Thomson was indifferent to formal religious beliefs, an analogous attitude was expressed in his scientific beliefs.

These beliefs were firmly held all his life and were well set by the time he was seventeen. The core of Thomson's originality consisted of a synthesis of imagination, combining the mathematical with the physical in the ways previously described. By the time he entered Cambridge University in 1841, his way of thinking was determined. He had already been introduced to Faraday's ideas; "An Essay on the Figure of the Earth" as well as "On the Uniform Motion of Heat" embodied that thinking. Interweaving all these separate ideas was Fourier's mathematics, "Fourier, the great French creative mathematician who founded the mathematical theory of the conduction of heat."[34]

CHAPTER THREE

Cambridge: An English Education

Had William Thomson gone no further than his father's coaching and the education he received at the University of Glasgow, he might still have made a creditable contribution to nineteenth-century science. At seventeen, Thomson's background was comparable to that of George Boole or George Green, who made original contributions to mathematics and science without benefit of a university education.[1] The characteristic of individualism fostered by a narrow education is that the work is idiosyncratic and finds its way into the mainstream of scientific thought after a slow process of recognition. Going to Cambridge was an experience that expanded Thomson's influence from that of an original thinker to one of the leading figures in the British school of science.

In his father's view, the Cambridge degree was necessary for Thomson's future candidacy for a professorship at Glasgow. Dr. Thomson decided that St. Peter's College (Peterhouse) was the best college at Cambridge for his son, and he wanted William Hopkins to coach him. In October 1841 father and son arrived in Cambridge by carriage. Dr. Thomson carried letters of introduction to smooth the way. One was to James Challis, Plumian Professor of Astronomy and Experimental Philosophy, another to Duncan Farquharson Gregory, editor and cofounder of the *Cambridge Mathematical Journal,* and the third letter was to William Hopkins. The three men had each made a contribution to the new British school of mathematics and each, in some way, could be helpful to a young man wanting to make his way in the world of mathematics.

Before Dr. Thomson left Cambridge, he saw his son settled in comfortable rooms consisting of a parlor, bedroom, and gyp's room. (The gyp was a college servant who kept the rooms in order and set the table for breakfast and tea.) The contrast with Dr. Thomson's college experience was manifest at every turn that day. James Thomson's trip of uncertain duration had brought him to the University of Glasgow with its blackened walls situated in an unhealthy slum. This time, father and son arrived on a scheduled stage and went to rooms reserved for William. They walked around bucolic Cambridge and along the beautiful Cam that passed along the "backs" of the colleges.

Through correspondence, Dr. Thomson maintained a dominant role in his son's life. His first letter urged him to bear the separation from home with equanimity. William was asked to write how he felt and how he was getting on "without concealment." The first step in William's growing independence was granted because his father had no experience in the ways of young gentlemen at Cambridge, but he did know about economy. "At the same time," he wrote, *always making moral correctness and propriety your aim above all things else,* you must keep up a gentlemanly appearance and live like others, keeping, however, rather behind than in advance."[2] Concern for his son's welfare was more intense during the first few months when such advice as "take care of yourself as to body, mind, and conduct" was appended to almost every letter.

In his letters, Dr. Thomson emphasized "the vast importance of . . . acquiring *accurate business habits* along with whatever else [one] may learn." Finances were a frequent topic in their letters. He required William to give a detailed account of his expenditures and then scrutinized the items. Did William require two pairs of shoes? According to the information in a booklet titled "Reform your Taylor's Bill," the coat he bought cost too much. William accepted his father's advice with equanimity, at times expressing slight exasperation. He wrote begging letters and was duly grateful on receiving money. William did reform his business habits. Where in 1841 he was offhandedly dismissing a two-shilling shortage in his account as being spent on something or other, by 1842 his accounts were correct to the halfpenny, but he was allowed some discretion on account of there being at Cambridge, as he wrote to his father, "an immense number of expenses which it is quite impossible to avoid, but which a person not accustomed to the methods of proceeding here may consider quite unnecessary." His father accepted these exceptions, and William exercised economy in many ways.[3]

Thomson made friends easily, beginning with those like himself who had just come up from the University of Glasgow. There were wine parties, breakfasts in each other's rooms, and long walks which served as breaks from the long hours of study. Conversation ranged from small talk about topics of the day to the main topic, the competitive honors examinations, or Tripos, that came at the end of the course of study. Every friend was a competitor as well. Arthur Scratchley, his best friend during the first year, appeared to be William's closest rival for a fellowship at Peterhouse. They cooperated on a routine that included seven and a half hours of study and three hours of exercise, including walking, boating, and time spent at a gymnasium, but the worry over the competition took William to his tutor to find out what he had to do to beat Scratchley. The two friends discussed the rising competition and because he feared it might become destructive, Scratchley migrated to another college.[4]

Thomson's main object in coming to Cambridge was the attainment of an honors degree in mathematics in order to support his candidacy when he applied for a position. At the end of his course of study, he faced the Tripos, an examination that lasted for eight days and was taken in two parts. During the first three days, candidates were examined on the basic principles of mathematics. After the first part of the exam, there was an interval of eight days for marking the papers. At the end of that time, a list was made of those eligible to take the final part, which lasted five days. The outcome of part two was published, and the candidates were listed in three classes. The highest category was "wrangler," an odd designation that was a vestige of the time when examinations were disputations on theological subjects and the best disputants, or wranglers, were selected. The man with the best set of papers was called senior wrangler, and the list continued as second wrangler, third wrangler, and so on. The senior wrangler was the man best drilled in writing solutions to mathematical problems. The number of solutions completed weighed more than the quality of the answers. Coaches drilled their charges to the point of having them practice with quill pens to improve their writing speed. The mathematical honors candidate also had to demonstrate a competence in classics and theology.

The best coach at Cambridge was William Hopkins, known as "the senior wrangler maker." Many students wanted to study with him, and Thomson was one of the few selected. Knowing of Hopkins's reputation, Dr. Thomson had chosen him for his son's coach without realizing that Hopkins had another side to his teaching that was to be very important to William's career. Hopkins disapproved of the method of merely drilling students in preparation for the Tripos. The fact that the coach was so dependent on pupils for his income too often meant, he wrote, that the tutor tried to supply the students' "imaginary wants and to relieve them too much from the toil, and thus to deprive them of the wholesome discipline of thinking for themselves."[5] Recognizing that there was no inducement or opportunity for study beyond the B.A.—there were scarcely a dozen situations in England for which a high degree of mathematical attainment was essential—Hopkins counted on undergraduate teaching to maintain "the scientific character of the country." In a booklet published in 1841, Hopkins stated that the best of the Cambridge students had by their training been conducted "to an eminence from which they might contemplate the fields of original research before them. . . ."[6]

While Hopkins approved of the definiteness of the examination questions as a means of correcting a tendency in students to be vague and indeterminate in expressing themselves, he did not think the examinations "encourage or finally test sufficiently the combination of those elements which constitute at once the unity and comprehensiveness of [their] knowledge, and which is the result of [their] last and highest mental efforts." Hopkins appreciated the unusual quality of Thomson's mathematical abilities and tried to correct what he saw was Thomson's "want of *arrangement* and *method*." Thus while Hopkins did not hesitate to drill his students, he made certain that they went further—to eminence. William informed his father that Hopkins did not want him to spend too much time drilling. When Hopkins showed his students a particular solution to a problem, he told them they need not go further than the point indicated, but then he proceeded to show them the possibilities for further analysis.[7]

Hopkins's reputation attracted the best talents. After selecting the most promising students, he knew how to orchestrate their mixture of self doubt and assurance to bring them to their peak for the examination. Every one of his actions seemed calculated, step by step, to tune his charges to the right pitch for the examination. An invitation to a party at Hopkins's house was flattering to entering students, indicating that the great man might take them as students. It was a tense time for all. Thomson noted that the crowd consisted of young men who looked like wrangler material. Who would be chosen? It was encouraging to know that Hopkins had been impressed enough by the students' references to invite them to the party, but the number attending exceeded the number that Hopkins was able to coach at one time. Consequently, William and his friends realized this was the first step in culling. Apparently, Hopkins had made up his mind about some of the students before that evening, and Thomson was one of them. Although Hopkins did not begin tutoring until after the students had successfully completed their introductory courses, he did advise them on what to study. After receiving an assignment, Thomson was told it was a good plan to remain at college over the Christmas holiday and study. Hopkins made his demands on his charges and those who would not or could not meet them were dropped. He advised Thomson which mathematics books to read and from time to time examined him to determine his progress; about two months later he thought it best that Thomson have a tutor. An undergraduate who was close to taking his degree was chosen, and he drilled Thomson on the low or beginning subjects. Hopkins himself managed the final assault on the Tripos.[8]

Hopkins taught small groups of students—about five at a time—giving each one the individual attention he needed. These classes were another step in the tension mounting toward the final examinations. Students like Thomson, aware of their own abilities yet sensitive to their deficiencies, responded in an expected manner. Thomson's goal was senior wrangler, and he quickly recognized his closest competition. He was W. F. L. Fischer who seemed to know everything about every subject and still considered himself not sufficiently prepared. "I must read very hard," Thomson wrote in his diary, "and try at least to be as well prepared as he is."[9]

The Tripos examinations had many of the characteristics of a thoroughbred horse race. Trainers or coaches built reputations on the past performance of their charges. Colleges were much like stables, which had standing according to the number of winners they had over the years. Good stock or likely candidates were identified by practiced eyes and groomed for the big event. Word spread through the colleges about who were the favorites. Betting was common. Like thoroughbreds, candidates were given special treatment and rigorous training that increased in intensity as the day for the contest drew closer.

Cookson, William's college tutor, informed Dr. Thomson that with steady reading and hard work his son would stand high in the examinations. As his studies advanced, William gained more confidence but was never allowed to become overconfident. When Hopkins's small group of potential wranglers began the study of mechanics, Thomson felt "less violent in my apprehensions regarding Fischer." Hopkins's choice of candidates and his success in training them made 1845 an unusually good year. In Thomson's class, he had three of the first five wranglers. Hopkins took direct and complete charge of them in 1843. They met alternate days with their coach and were given examination questions; on the other days, they met individually with him to discuss their solutions. In spite of the rivalry, or perhaps on account of it, the students became close friends. Thomson, Fischer, and Hugh Blackburn spent hours together, studying, walking, and swimming.

During the summer of 1844, Hopkins took his group with him to his summer holiday at Cromer by the sea. The relationship between the students was very much like siblings, including the characteristic of sibling rivalry. Thomson confided to his diary in February 1843: "Went to Hopkins at 10 and was maliciously glad to find that Fischer had not done all the problems. Blackburn had got solutions for all, but nobody had given interpretations except myself." By May of that year, Hopkins had chosen his favorite student. When he was alone with Thomson, he told him he was certain to become senior wrangler.[10] How adroit Hopkins was in the management of his charges. By constantly drilling them in the subjects of the examinations and with just the right amount of encouragement, he inspired his students. Another aspect to Hopkins's teaching that distinguished him from the other coaches was his concern that his students, after doing well in the Tripos, would go on to do original research.

Hopkins was William Thomson's link to the Cambridge school of mathematics. While Hopkins was schooling his students in the routine mathematics they would be called upon to do in the Tripos examinations, he encouraged them to think along the newest lines of mathematical analysis. These views consisted of using the principles of algebra in solving problems on the calculus. Integral or differential calculus was used to solve physical problems dealing with specific states, such as the amount of matter in an irregularly shaped solid or the attraction of a solid on a point. A differential equation describes changing conditions which occur when heat passes through a solid or an electric current flows through a circuit. The combination of algebra and calculus was known as analysis. Although analysis was commonplace as a study on the Continent, in the 1840s it was considered too novel by some to teach to Cambridge undergraduates. Hopkins was using a book written by George Peacock, one of the founders of the new Cambridge school of mathematics.[11]

Peacock along with Charles Babbage and John Frederick William Herschel had undertaken to reform British mathematics and to introduce the system used on the Continent. In 1812, the three began meeting informally for Sunday breakfast and styled themselves The Analytical Society. By 1823, they succeeded in introducing the new mathematics to Cambridge in the most effective way possible—by proposing questions for the mathematical Tripos.[12] Peacock was able to exert this influence since he was a fellow of Trinity College and was selected to help write the Tripos questions.

Thomson first learned of the new mathematics indirectly when doing research for his "Essay on the Figure of the Earth" at Glasgow. One of his references was Airy's tract. Airy was a student of Peacock who had written his tract to demonstrate the effectiveness of the new system. In addition to Hopkins's espousal of the new system at Cambridge, Thomson found that D. F. Gregory, editor of the *Cambridge Mathematical Journal,* was a proponent of Peacock and the Analytical School.[13] Gregory was one of the founders of the *Cambridge Mathematical Journal* and became its first editor in 1837. The preface to the first issue of the *Journal* referred to Cambridge as the new center of mathematical studies in Britain. The object of the *Journal* was to encourage the writing of original papers by Cambridge men and to help spread ideas from the Continent by publishing abstracts of important articles that appeared in foreign journals.[14]

Gregory began work on a revision of the original work written by the three organizers of the Analytical Society, but he found that so many changes had been made in mathematics during that brief span of time that a new book was warranted. In 1841, he published *Collection of Examples of the Processes of the Differential and Integral Calculus.* It was meant to take the place of Peacock's *Examples,* which was out of print. Thomson informed his father that the new book was much better than Peacock's and was causing a stir in Cambridge. It was credited with bringing the calculus of differential operators into common use.[15]

Before he came to the university, Thomson had written three articles for the *Cambridge Mathematical Journal.* Consequently, when he came to Cambridge he had conversations with Gregory about difficulties that had arisen over the third article, "On the Uniform Motion of Heat." Gregory befriended Thomson and encouraged him in his original research as well as discussing his own work. Gregory was writing a new book on three dimensional geometry from the analytical viewpoint. He contended that there was an analogy amounting to an identity between the operation of symbols in algebra and the differential calculus which suggested solutions to thus far insoluble differential equations. Thomson and Gregory struggled together over these new ideas, and Thomson wrote to his father that Gregory was "not however quite clear about the principles" of his discovery.[16] Gregory never finished the work. He died in February 1844 at the age of thirty-one.

Was it not harmful for Thomson, who was priming for the Tripos examinations, to be doing original research and writing articles for the *Cambridge Mathematical Journal?* His college tutor wrote to Thomson's father questioning the propriety of writing any more articles before he took his examinations. Archibald Smith, fellow of Trinity College and senior wrangler in 1836, thought otherwise. He had met Thomson at Cambridge and now interceded on his behalf with Dr. Thomson, who wrote that Smith "thinks you should do *little*, but that you should not make yourself a mere bookman, but that whether you publish or not, you ought to think for yourself, and, at least, to accumulate materials that may be used hereafter."

Dr. Thomson thought that a very wise view since he did not have the power to stop his son from thinking for himself and developing new ideas.[17] Thomson was in the midst of the Cambridge school of mathematics led by Gregory when it was most active, and Thomson was being influenced by what influenced them—the Continental mathematicians. Mathematical theory was undergoing important changes through the method of combining geometry, calculus, and algebra. The resulting operations were used for thinking about forces of attraction: gravity, electricity, and magnetism. The subject suited Thomson perfectly.

The more intense the training for Tripos, the more active was Thomson in original work. During his three and a half years as a Cambridge undergraduate, he wrote a total of ten articles. Four of these were written in 1843, after Hopkins began direct supervision of William's studies. An article titled "Propositions in the Theory of Attraction" appeared in the November 1842 and February 1843 issues of the *Cambridge Mathematical Journal*. The article has a reference to George Green's term "potential," which Thomson had read about in a secondary source.[18] The concept was derived from Newton's theory of gravitational attraction. Newton calculated that force by multiplying the product of the masses of two attracting bodies and dividing by the square of the distance between them. Green calculated the forces of electrical attraction and repulsion by summing up the elements of imaginary mass of the attracting body and dividing by a function of the distance between the attracting body and the attracted point. The concept interpreted force from the point of view of the calculus, and it was one that fitted a geometrical conceptualization. Points of equal or constant potential were located on surfaces called "surfaces of equilibrium." Thinking of force as a locus in space meant that even though the ultimate cause of the force was unknown, calculation of the magnitude and direction of the force was possible. Whether the source of the force was a material substance or not was controversial and according to this method was of no consequence. The *effect* of the force was as if it were a substance. One might imagine it as due to a surface pressing against the point being acted upon. Once the effect was conceptualized as a surface, the calculus of mass as well as the theorems of solid geometry were applicable. By using the methods of solution of either analysis, the combination of calculus and algebra, or those of solid geometry interchangeably, many complex problems arising from the consideration of heat, electricity, and magnetism were now soluble. The key unlocked many formerly unrelated puzzles.

The May 1843 issue of the *Cambridge Mathematical Journal* carried two of Thomson's articles, "On the Attraction of Conducting and Non-Conducting Electrified Bodies" and a "Note on Orthogonal Isothermal Surfaces." The ideas were, as Thomson thought of them, related. The surfaces of equilibrium, which described areas where electrical force was constant, were analogous to surfaces of constant temperature, that is, isothermal surfaces. These surfaces were more easily managed by the mathematical methods invented by Gregory and others; furthermore, they served as a way of thinking about the phenomena. Thomson and the other mathematicians regarded this system of dealing with forces as valid since all reasoning was guided by mathematical logic once the geometry was determined. It was admittedly imaginary, but the exactness of the system was assured by the use of mathematical logic and the fact that results agreed with experimental results.

The conviction that the imaginative use of geometric and mathematical symbols was an accurate way of treating heat, electricity, and magnetism was a narrow one; the mathematician's certainty did not allow for other imaginative ways of treating the subject. Being among mathematicians, Thomson thought like them. When he and Gregory discussed Faraday's latest research, the experimentalist's way of reporting his results was harshly "abused," Thomson reported. Reading about Faraday the next day, Thomson confided to his diary, "I have been very much disgusted with his way of *speaking* of the phenomena, for his theory can be called nothing else."[19] He can have meant only that Faraday's description of his work, that is, using such imaginary physical concepts as lines of force, was merely "*speaking*" about the forces and not dealing with them analytically. Soon after, Thomson found a way of describing Faraday's lines of force mathematically and felt less antagonistic toward Faraday's way of thinking.

Gregory encouraged Thomson to write articles for his journal, and conversations with other mathematicians at Cambridge, such as William Walton, Ellis, and Challis, as well as his fellow students Fischer and Blackburn, were inducements to write. French mathematicians were thinking along the same lines, and they had to be read for new ideas as well as assurance that one was not anticipated. Thomson started a subscription to Liouville's *Journal de mathématiques* and found that an article by Bertrand on orthogonal isothermal surfaces used almost the same words he had used in his article printed in the May and November 1843 issues of the *Cambridge Mathematical Journal*. He had no time to follow up with a claim of originality, for at that moment he was engaged in examinations and his studies. A short time later, his "Note sur la théorie de l'attraction" was inserted in Liouville's *Journal*.[20]

To the satisfaction of Hopkins, Thomson resolved the conflict between the necessity to cram and his wanting to do original work but often felt chafed by the interference with his original research. Mathematical diary entries were often terminated because he had to resume his studying. In the midst of concentrating on his studies, Thomson's imagination brought new ideas to the surface of his consciousness. These appeared as dreamlike musings which he described in his diary:

> Yesterday night I got foul of the orthogonal surfaces again, and sat till 12½ with my feet on the fender, but got no satisfaction. Today after coming from Hopkins I have got some new ideas, but not the ones I wanted.[21]

Thomson's thinking progressed as rapidly and as far as that stage in his development allowed. The Cambridge years were a time when he absorbed the new mathematical ideas of the Cambridge school and the Continental mathematicians. If he had continued at the same level, he would have become an important member of the Cambridge school. However, he was not to be content with that. But the extension of his thought entailed a synthesis of physical and mathematical conceptualizations. He was not able to make that synthesis before 1845.

A conflict which affected Thomson while at Cambridge was the dispute between the geometers and those, like Hopkins, who wanted to teach the students the new method of analysis. During the period in which Thomson was a student, 1841–1845, analysis, the combination of algebra and calculus, prevailed, but the geometrical view was a strong undercurrent. William Whewell, master of Trinity College and an advocate of a Cambridge education as a liberal one, believed geometry was the only form of mathematics for disciplining the minds of future lawyers and clerics. Students, he thought, did not need to know and were not ready to learn the applications of mathematics to physical subjects. Geometry, not Gregory's new ideas for applying the principles of algebra to calculus, was a necessary part of a good education. When a student learned geometry, he was "rendered familiar with the most perfect examples of strict inference." Teaching students the new mathematical ideas led him, Whewell wrote, to "the conceit of a supposed knowledge of something which his predecessors did not know."[22]

During the years Thomson was at Cambridge, Whewell represented a minority view. The questions for the Tripos examinations were written by followers of the Analytical School, but his undercurrent of dissent was strong enough to be given a sympathetic hearing when a special royal commission was convened in 1850 to investigate education at Cambridge. The commission's report reflected the growing disfavor that was directed at Cambridge and Oxford. The *Edinburgh Review* complained that the universities were not providing professional training for solicitors, medical practitioners, engineers, bankers, or merchants who were badly needed for a rapid commercial and industrial expansion.[23]

However, while there were demands being made for a more specialized professional education to support an expanding technology, Cambridge was harboring the new mathematical school that supplied the theoretical base for British industrial expansion. Analysis, which people like Whewell were complaining about, was the method used by William Thomson to develop the theory that made the Atlantic cable possible. Britain's world domination of submarine cables during the latter part of the century was founded on that theory. Other countries such as the United States were turning to more practical, applied engineering education, but it was a British engineer who made the original design for the electric generators at the Niagara Falls power station.[24]

Although Thomson published articles and prepared for the mathematical honors degree, he was not yet certain he was going to be a mathematician. Most of the young men at Cambridge planned to make their professions in the church or at the bar. Thomson never considered taking holy orders, and for a time, it seemed that the law was all that was available to him. He visited the courts in town and was impressed with the proceedings; he decided that he "could reconcile [himself] to the bar, though it would be a great shock to [his] feelings at present, to have to make up [his] mind to cut mathematics."[25] Thomson might have made some contributions to mathematics as Archibald Smith had done. Smith was elected to the Royal

Society and contributed original papers on mathematics while maintaining an active law practice. But Thomson's interest in mathematics was too intense to allow merely a part-time interest.

Thomson's shy demeanor belied an intensity of feeling that was revealed in his diary and was the cause of a personal crisis in 1843. He was no more sentimental than the average youth his age. However, his attempts to repress these feelings makes one curious. "I have not been at all sentimental, prospectively or retrospectively (with slight exceptions at least) today. I have been working at electricity, and have got some new ideas. (see journal)" The reason for taking his sentimental pulse was that he had been taken unawares the day before by some reminiscing. "This morning I got hold of my *math journal,* and spent an hour at least in recollections. I had far the most associations connected with the winter in which I attended the natural philosophy [class] and the summer we were in Germany. I have been thinking that my mind was more active then than it has been ever since, and have been wishing most intensely that the 1st of May 1840 would return. I then commenced reading Fourier, and had the prospect of the tour in Germany before me."[26]

The cornopean, or cornet, began as a diversion, but eventually the tedium of his mathematical training made it an earnestly pursued hobby. "I must not," Thomson wrote to himself, "let the cornopean take up so much time in the future." In spite of his efforts, he often slipped in that resolution. Finding music open on his mantle, he frequently wasted evenings that he wanted to spend with mathematics. At another time, he came to his rooms "determined to read till hall, and started by playing the Easter hymn for half an hour." Thomson was one of the original members of the Cambridge University Musical Society and for a time was its president, although music for him was never more than a mathematician's hobby.

One April evening William wrote: "This is a beautiful moonlight night, and my rooms are quite romantic. If I were only sentimental enough to enjoy it, I might lose a great deal of time looking out of the window." And another moonlight night he wrote:

> After I have got everything ready to go to bed, I have been looking out of the window, and have got back my journal, to endeavour to fix my impressions. The moon is shining brightly on the mist which lies in the meadow . . . like the silvery clouds we saw from Ben Lomand. I have been looking out of the window for a long time, and listening to the distant rushing of waters, the barking of dogs, and the crowing of cocks.

He had no time for these musings. On another occasion, he had "a long train of associations which I have not time to recount."

Thomson had another weakness which was slight at first and became absorbing in 1843. That was his pleasure in reading fiction. With prize money won for excellence in mathematics during his first year at Cambridge, William bought a copy of Shakespeare, arousing the ire of his father: "It is so like Dr. Nichol: and see what the like of this has brought on him." Undeterred, William kept the book. On a number of petty issues like these, William gradually began to assert his independence from his father. He never disobeyed him, but the distance from home and his father's unfamiliarity with what was expected at Cambridge loosened the bonds. There were rare occasions when Dr. Thomson was thwarted in his attempt to guide his son's behavior. One such crisis occurred over William's brief period of rowing madness.

William's rigorous schedule of study was broken only by periods devoted to exercise. Walking for a time sufficed. However, he found the countryside boring in comparison to Scotland, and he became interested in rowing. Without consulting his father, he bought half interest in a "funney," a one-seater boat. He now exercised as much as he wanted and did not have to join a boat club where the rowing crowd congregated. The detailed explanation for his act included the confident statement that his father "would not be displeased."[27] He was wrong.

His father's reply upbraided him for being a foolish young man. First of all, William had not consulted his father, and he obviously had been hustled into making a poor bargain. Dr. Thomson did not believe that William would pay "Seven pounds for a tub that will only hold one person!!!" Instead of wilting, William wrote a long letter defending his purchase but ended by offering to sell his half of the boat if his father insisted. William's college tutor was consulted, and he reported that William had acquired good companions who were not likely to lead him astray. Elizabeth informed William that "though papa was at first quite displeased with your purchase he is now I think quite reconciled to it and will not disapprove of your rowing for exercise."[28] There was more at stake than the purchase of the funney, for the ties through which his father directed his life began to loosen. Those ties were never broken during his father's lifetime, but William's spirit required more freedom than his father was willing to give him. Consequently, the son had to make stands to win some degree of freedom. During this time, William was contending with the personality conflicts that come with maturity.

Dr. Thomson was willing to recognize his son's maturity in mathematical matters, and many letters were taken up with discussions about mathematics. William gave his father suggestions for examination questions and supplied alternative solutions to help him in the grading. Thomson wrote home about the publication of the latest mathematics books and made suggestions for purchases by the University of Glasgow Library. All this advice was sought by Dr. Thomson and readily followed. The contrast between the correspondence on mathematical matters, which they conducted as colleagues, and the continued paternalistic care with which Dr. Thomson supervised William's personal affairs reflects the conflict that William was forced to face. How does a lively, sensitive young man resolve the conflicts between his emotions and the mature, rational side of his nature that was engaged in original mathematical work?

Was Thomson one of those "dry-headed, calculating, angular little men" that Alfred Tennyson saw when he was at Cambridge?[29] Tennyson's "dry-headed" young men were those, unlike himself, who submitted themselves to drill in preparation for mathematical honors degrees. Does the description fit Thomson, who did well in his cramming? Certainly not. He was not that type when he came to Cambridge. Was he in danger of becoming the heartless, calculating mind that Tennyson detested? He might have, had he submitted completely to the Cambridge type of training.

When Thomson came to Cambridge, he was a mixture of the angular rationalistic spirit and the emotional. Interest in the cornopean derived from his sense of symmetry and order, that is, a sense which attracts mathematicians to music for its harmony more than for its emotional or aesthetic content.

Even when Thomson spoke in the privacy of his diary of being sentimental, the manifestations of his aesthetic sense made him uncomfortable. The exhilaration he felt in viewing the world through the means of mathematical logic created a conflict within him. At the ecstatic level of intensity, the aesthetic and the logical senses conflict. At lower intensities, those who find satisfaction in mathematics as an avocation can find equal satisfaction in painting or literature. At the creative level, there is no dualism: painters may become musicians or mathematicians may move to the logic of philosophy, but there is no crossover route between the logical and the aesthetic for creative people. If William Thomson wanted to be completely absorbed in mathematics, he was going to have to sublimate the aesthetic sense that was distracting him.

Literature was attracting him at the time, and he was responding emotionally. Evidently, he began with a strong resolve not to be diverted, for he wrote to his father: "I see . . . that you are afraid of me being led away by my love for lighter literature, but you need not be in the least afraid of that, as I very rarely either feel inclined for reading, or read, anything but mathematics of some kind." Nevertheless, he was decidedly being led away in 1843 when he discovered that Hugh Blackburn, his close friend and competitor in Hopkins's class, was "quietly improving his mind, in various ways." Blackburn was taking enough time off from mathematics to read Shakespeare, Beaumont and Fletcher, and Ben Jonson. If Blackburn—whose goal was also to be a high wrangler—could indulge his tastes in lighter literature, the indulgence being called improving the mind, Thomson found license to do the same. At the time when Thomson's eyes and mind were straying out his window to the beautiful scenery, he was also staying up too late reading *Henry IV*. He sampled Shakespeare's poems and found them very much to his liking. The compelling emotion took him so far as to begin reading *Evelina*, a romantic novel by Fanny Burney, first published in 1778. Losing all track of time, he read until daylight. After putting the book down, William "spent a long time looking at the sheep, and listening [to] the birds, whose singing *filled* the air."[30]

The strongest impression on Thomson's sensibilities during that demonic year, 1843, was made by reading Goethe's *Wilhelm Meister*. Fischer, of the awesome mathematical prowess, introduced Thomson to the work that was almost singlehandedly fostering a romantic movement among German youth. That fall, Thomson and Blackburn were reading *Wilhelm Meister* nearly every evening, and it was having a most disturbing effect on Thomson. Part of his unease was recorded in his diary on Monday, 23 October 1843.

> During last week I have been rather unsettled and not applied myself to reading nearly as much as I could have wished. [He was referring to his mathematical studies.] The idleness however did not depend upon external circumstances, as I have been in my rooms almost every night at or before 7, but partly to my having Wilhelm Meister. . . .

The rest of the page is torn off.[31]

The conflict between his aesthetic and rational sense had reached its climax. Goethe's romantic novel had aroused feelings which William found disturbing and uncomfortable. Interestingly, in 1850, when James Clerk Maxwell was a first-year student at Cambridge, his experience was such that he said *Wilhelm Meister* should be "read with discretion." Thomson hastily abandoned improving his mind in that direction, put away the fiction, and concentrated on his mathematics. The redirected energy produced three published articles concerning the mathematics of the physical imagery of forces of attraction and orthogonal isothermal planes.[32] *Wilhelm Meister* was not an experience he wanted repeated. It was unlike the ecstatic sensation he recalled when first reading Fourier.

In 1843, William Thomson had another struggle with an interest that threatened to divert him from his studies. It was boat racing. When the university boat races began in early March, Thomson went to the Cam to watch them. Sculling in his one-seater was simply a form of exercise, but now Thomson saw racing as an exciting sport. He found, perhaps to his surprise, that he felt strongly about the Peterhouse boat's showing. Among the cheering, rushing crowd along the banks, William felt dejection when Peterhouse was bumped by Trinity and an unexpected elation when the Peterhouse boat succeeded in "keeping away." Thomson went to the races several times; when he heard that a new crew was being formed, he was "very much tempted to join, but both on account of reading, and of what my father would think of me joining, I shall delay." Since he did not intend to resist the temptation altogether, he was determined to join a crew sometime before taking his degree. A diary entry written at 2:30 A.M. read: "I have been thinking on the boat more than anything else, all day."[33]

By the middle of April 1843, Thomson was rowing with the crew during practice and writing that he was dissatisfied with his physical condition. He was "trying to train furiously." Toward the end of April, his diary was filled with comments about boat racing—he now pulled number seven position in the Peterhouse first boat. By 1 May, the excitement seemed to be wearing off. In the middle of May, the boat racing had commenced in earnest. The Peterhouse boat was in first position, at the head of the river, and although the odds were that it would be overtaken, "[the crew] astonished the university by keeping away." Thomson was ebullient. "We had a glorious pull for it, and I shall remember for my whole life, the work of 7 minutes last night. My pleasure at keeping away was beyond anything I have ever felt."[34]

The next autumn, Thomson raced in single-seater boats or sculls, but recorded that his "desire for racing was however whetted." His father was annoyed that his racing fever had continued and thought of it as being too much like gambling. Thomson won the Colquhoun sculls in a competition of fourteen boats. This gave him the honor of holding a pair of silver sculls for one year. His diary ended with a last word about boating: "I need not stop to commemorate anything about boating last term [spring 1843] or this, as I suppose I shan't forget it in a hurry; I shall at least remember everything worth remembering about it."[35]

The ecstasy was limited to seven minutes in boat racing; there was no higher goal after he had won the silver sculls. It compared not at all with the exhilaration of pursuing a line of inquiry that promised greater and greater truths or with the knowledge that there was no limit to aspirations because there was no ultimate truth obtainable by man.

Thomson's experiences at Cambridge were a series of choices presented to him. His response, true to his personality, was the rational over the romantic. During this period of rising aspirations, Thomson felt an intensified psychological disturbance. Once his choice was made, his determination crystallized. For the rest of his life, an extraordinary degree of concentration was spent on original scientific thought. He wrote no more personal diaries to chart his emotional responses. Nor did the romantic literature of Shakespeare and Goethe hold a fascination for him. Thomson's constant companion throughout his life was his mathematical notebook.

CHAPTER FOUR

A European Reputation

The subjects for examination are pressing in on all sides and seem very formidable at present but the greater the pressure is now, the greater will be the relief this day four weeks when all will be over. I do not feel at all confident about the result, but I am keeping myself as cool as possible, and I think I shall not be very much excited about it when the time comes. One thing at least I am sure of is that if I am lower than people expect me I shall not distress myself about it, and if any of you lose money on me I shall consider it your own fault for giving odds.[1]

Expectations for William Thomson becoming senior wrangler were running high. After all, by the time of the Senate house examinations in 1845, he had published eleven articles in the *Cambridge Mathematical Journal* and one in Liouville's *Journal de mathématiques.*[2] Thomson was regarded with a feeling close to awe at Cambridge. Robert Leslie Ellis, one of the examiners in Thomson's year, told another examiner, "You and I are just about fit to mend his pens."[3]

One can easily imagine the consternation of friends and the disappointment of Dr. Thomson when the results of the Senate house examination for the bachelor's degree with mathematical honors were announced in Cambridge on 17 January 1845. William Thomson was *second* wrangler. Many years later, when Lord Kelvin was reminded of the Tripos, he recalled how he had allowed himself to be distracted and thus missed becoming senior wrangler:

One paper was really a paper that I ought to have walked through, but I did very badly by my bad generalship, and must have got hardly any marks. I spent nearly all my time on one particular problem that interested me, about a spinning-top being let fall on to a rigid plane—a very simple problem if I had tackled it in the right way. But I got involved and lost time on it. . . . [4]

William Hopkins, Thomson's coach, wrote a letter to Dr. Thomson immediately after the examination results were posted. Hopkins found William's placing understandable in the light of the type of testing. The shortcoming in Thomson's thinking, as Hopkins saw when he first began tutoring the young man, was the lack of order in writing out his solutions. Although this defect had been remedied partially during his training at Cambridge, especially under Hopkins, Thomson was still not conditioned to answer the mathematical questions in the Senate house in a way that demonstrated his competence. While others were "simply *answering a question,*" Hopkins noted, Thomson was "writing a *dissertation.*" However explicable the results were, Hopkins did not doubt Thomson's first place among the Cambridge graduates in "high philosophic character both of mind and knowledge." What was of more lasting importance, Hopkins assured the older Thomson, was the "expansiveness of view, which is so eminently characteristic of your son, and affords so sure an indication of future distinction."

One of the Senate house examiners told Hopkins that, while the senior wrangler of 1845 might not ever be known beyond the walls of the university, William Thomson was building a *"European* reputation" for himself. For some people, Hopkins wrote, academic honors were the culminating points in their careers. Hopkins had known a few young men for whom far more valuable honors came after they left the university. Over the years, Hopkins had learned to recognize special talent and saw an unusually bright scientific future for William Thomson. *"That prediction I have never uttered more confidently than I now make it with respect to your son.* I have never been mistaken yet, and if I prove so now I'll never prophesy again."[5]

It is difficult to determine exactly what was meant by a European reputation. When William's father suggested to a colleague, Dr. William Thomson, that William would be a likely candidate to fill the anticipated vacancy in the chair of natural philosophy at Glasgow, his colleague, professor of *materia medica* at Glasgow, was at first struck by William's youth and then asked about his *"experimental* acquirements, particularly in chemistry, and . . . mentioned [James D.] Forbes as being *in this respect* one of the first men of the day, and as being of 'European reputation.' "[6] A skill at performing demonstration experiments and giving popular lectures was valued in a candidate who had to bring science to the level of the students at Glasgow. At Cambridge, having a European reputation meant being known on the Continent through the publication of mathematical articles. Thomson had adopted more of the Cambridge outlook. As he wrote to his father on the day the results of the Tripos were given, he was not disappointed in the outcome from "a mathematical point of view" and hoped his father did not consider the time at Cambridge misspent. "I feel quite satisfied," William wrote on that bleak January day, "that I have spent as much time on reading and preparation as I could consistently with higher views in science."[7] While waiting for the results of the examinations, Thomson had finished two papers and sent them to the *Cambridge Mathematical Journal.*

Dr. Thomson was not one to waste time with regrets. One can only guess at the depth of his disappointment on learning the examination results. The timing of events gave Dr. Thomson his first grandson, James Thomson Bottomley, born to William's sister Anna, now Mrs. William Bottomley, who was living in Dublin. Letters from his sisters and brothers began by voicing their chagrin at his placing second but lapsed quickly into the latest news about the Bottomley heir.

Aunt Agnes took the news about William not becoming senior wrangler the hardest. She wrote to Elizabeth: "I am most *desperately* disappointed that he has not taken the place that I am sure he is entitled to, and which I think he has only lost from scattering his exertions over too wide a range." The scattering of exertions was in her view the working on articles which were so important to her nephew in his higher views of science. "I am not consoled to learn that so and so, and so and so, stood second." She was referring to the well-known fact, expressed by Hopkins, that second wranglers often did better in later life than the senior wrangler. "I expected him to stand first, and the only thing that reconciles me is the conviction that *we* all needed this mortification."[8]

Besides his concern for Dr. Thomson's disappointment, William worried whether not being senior wrangler hurt his chances for the soon-to-be-vacant professorship of natural philosophy at Glasgow. He had been assured by friends at Cambridge that he would make a much better professor on account of his original research than if he had spent that time training to beat Parkinson (the senior wrangler). But that was the view held at Cambridge. Dr. Thomson had been quietly polling the faculty at Glasgow as to their views of his son's qualifications for the chair of natural philosophy. He had started preparations for the campaign in 1843, and the most serious objection was over William's inability to give popular lectures.[9]

There was dissatisfaction at the Scottish universities with the theoretical mathematical training given at Cambridge. This education produced men who had only an "X plus Y" knowledge of science. Graduates of Cambridge were too "high," they thought, to be able to reach the students at Glasgow. Thomson had been made aware of this criticism, and he suggested to his father that he go to Paris after finishing his studies at Cambridge. In France, he would have the opportunity to learn about popular lecturing by listening to some of the best-known popularizers of science. William's friend Fischer had attended lectures in Paris by Liouville, and since there were very few auditors, Fischer had become acquainted with Liouville. It was an exciting prospect for Thomson to think about, walking home from the lectures with Liouville and talking about the latest mathematical subjects.[10]

Thomson was unaware of the contradiction. Liouville was obviously *not* a popular lecturer or else his class would have consisted of hundreds, making it impossible to have a private conversation with him. Dr. Thomson replied that in the opinion of some at Glasgow, going to Paris to learn about popular lecturing would strengthen his candidacy considerably. Dr. Nichol, William's natural philosophy professor at Glasgow, strongly recommended the excursion, and Dr. Thomson was willing to approve it provided the *material* benefit justified the cost. As for William's idea of having the opportunity to talk about mathematical subjects with Liouville, that was "wormwood and gall."[11]

Before leaving Cambridge, Thomson had one more opportunity to demonstrate his scientific talents by competing in the Smith's prize examinations. Two of the prizes were open to new university graduates, and these examinations were on the "higher" subjects of mathematics and natural philosophy. Hopkins and others at Cambridge were proven right; Thomson was first in the Smith's prize examinations, and Parkinson, the senior wrangler, was second. Thomson's friend Ludwig Fischer wrote to Dr. Thomson commending his son:

> No one . . . who has the good luck of becoming acquainted with William . . . can help feeling deeply interested in a young man who devotes himself with ardour and commensurate success to science. But I have additional cause of feeling grateful to him as I have often and often received information in difficult parts of our studies from him and encouragement under the depressing drilling system which too much prevails at Cambridge, and to which alone I feel confident we must ascribe our disappointment in not seeing your son Sen. Wrangler, as he is undoubtedly the best mathematician not of this year only but of many years past. I hope his success in the Smith Prize Examination may have made you amends, as it certainly afforded a proof of his vast superiority.[12]

Dr. Thomson thought it best that William go to Paris, and as soon as the examinations were over he would begin to obtain letters of introduction. The cost of the trip was justified on two counts, and both of these were judged to improve William's chances for the natural philosophy chair at Glasgow. First, during his stay in Paris he was to meet as many important scientists as possible. William was told to make a list of the most important French scientists, and Dr. Thomson would obtain the necessary introductions. Second, William was to attend as many popular lectures on science as possible, learning the art of giving similar lectures and spending enough time in a laboratory to learn how to handle apparatus in order to be able to do the demonstrations for his own lectures.[13]

Soon after the Smith's prizes were announced, Thomson left for Paris with Hugh Blackburn, another student of Hopkins and Thomson's close friend. Blackburn was fifth wrangler that year, and Thomson must have convinced him that the trip to Paris was needed to prepare him for a university post. The Paris that the two young men saw had been made a showplace during the fifteen years of Louis Philippe's administration. The latest technology had been introduced—streets and footways were paved and sewers had been laid. The new gas lighting, much admired by foreigners, made Paris a city of light and the streets as lively by night as they were during the day.[14]

Prosperity caused a surge of middle-class culture in Paris. There were over three hundred *Cabinets de Lecture,* or reading rooms, where all the periodicals and journals of the day were found. Parisians felt they had to read several journals to hear all sides of a question. The improvement in living conditions based on new technology revived interest in science, as it always does. Public lectures on science were very popular. Arago was lecturing on astronomy at the royal observatory, and Baron Charles Dupin was speaking on mechanical philosophy. Attendance at both these series was between six and seven hundred people of all ages and of both sexes. As well as being a leading research scientist, Arago, "the French Faraday," gave eloquent popular lectures.[15]

The exhibition of a materialistic culture in Paris was detestable to Alexander Herzen, who arrived from Russia in 1847. Where Thomson had come to learn and bring back knowledge of advanced teaching in science, Herzen thought France was in need of redemption from the materialism technology had produced. He saw a different side of Paris by mixing in political and literary circles, as well as attending the theater, which he thought was corrupted by the bourgeoisie, or popular culture. Thomson did not take art that seriously, going only occasionally to the theater. In Paris, he found time to go to the *Opéra Comique* twice and went so far as to attend a masked ball. He delayed until last the task of delivering a letter of introduction from his Aunt Agnes. Thomson had promised one of Blackburn's brothers that he would have Blackburn present letters of introduction "to some very desirable people in the fashionable world." Blackburn refused to carry out the obligation. Despite Blackburn's unwillingness to fulfill family obligations, he and Thomson did have time for nonscientific enjoyments. They visited a friend's brother, an artist who lived in the Latin Quarter. With him, Blackburn and Thomson enjoyed small friendly dinners in the company of artists. Neither in his diary nor in his letters did Thomson make any mention of these visits.[16]

Blackburn and Thomson immediately set a schedule for attending lectures. They went to the Sorbonne to hear Pouillet's lectures on static electricity. Thomson took notes on the lectures, noting particularly those experiments used as demonstrations and the ones most appreciated by the very numerous and "popular" audience. They attended a number and variety of lectures in order to see as many different types of demonstration lectures as possible. They found that Dumas's lectures on chemistry were well illustrated with experiments, and they were impressed with the facility he had as a lecturer; everything was organized to cover a large number of experiments in a short period of time. At the Collège de France, they watched and listened to Regnault lecture on physics. There was some time for listening to lectures on the more advanced topics in science, but the two men had to be selective in these.[17]

Since the revolution, France was emphasizing useful knowledge. Schools such as the newly established Ecole polytechnique taught science as part of an engineering education. French scientists such as Ampère and Arago taught applied science at the polytechnique and also did their research in pure science.[18] At a time in which the upper class was moving toward a more practical and professional education, the middle class, through popular lectures, was learning more about the world of science in general. General education was going on in the streets, as it were, at the same time that the universities and technical institutes were becoming more specialized.

William Thomson's brief stay in Paris was not sufficient to change him into a popular expositor of science; furthermore, he was not so inclined. In Britain, it was a rarity for a research scientist to be able to draw large popular audiences the way Arago and Dupin did. Only Faraday, in 1845, had developed that talent, and he was an experimentalist, not a mathematician like his French counterparts. France

was, as a matter of fact, unique. In other countries, Britain and the United States, for example, popularization was a specialization not practiced by research scientists. Popularizers later in the century, including Tyndall, Huxley, and Spencer, were outside of the select scientific community of Thomson, Stokes, Tait, and Maxwell. They were also outside of the established universities of Oxford and Cambridge.

The University of Glasgow wanted a person who was well known to other scientists for his original research, one with a European reputation, and a popular lecturer. A well-known scientist would give status and a certain luster to the university, but it was *essential* that the occupant of the chair of natural philosophy be able to teach science to the level of Glasgow students. Dr. Thomson knew from experience what was wanted and sought the advice of other faculty members. He passed this advice on to William and directed his son's activities by mail. William was to write at least once a week, but he need not go over a quarter of an ounce per letter unless there was something remarkable to report. An important step to assuring the Glasgow faculty that Thomson had a European reputation was to have testimonials from "the great men of Paris."[19]

Thomson and Blackburn had letters of introduction from Cayley and Hopkins at Cambridge to some of the Paris notables. Dr. Thomson was collecting and forwarding to William a good number of introductions as well. He wanted William to meet the important French scientists and, if possible, to become well acquainted with them. Arago and Biot were high on Dr. Thomson's list, and he had in mind others who were well known enough to make an impression in Glasgow. William's father, having in mind at all times the letters of support a candidate needed for the natural philosophy chair, was as interested in having letters of introduction *from* important British scientists as he was in having William meet the important French scientists. Professors Kelland and Forbes of Edinburgh obliged. Sir David Brewster, at Dr. Thomson's behest, agreed to send a letter of introduction even though William was not known to him personally. Do not be surprised, William's father wrote to him, if you receive a letter of introduction from "no less a personage" than Lord Brougham.[20] That part of Dr. Thomson's campaign was going well. It was not so much that he was creating a reputation for his son; after all, William had done well at Glasgow and had distinguished himself at Cambridge. The raw material for a scientific reputation was there; the letters were a recognition of the achievements of a still very young man.

Two of William's friends on the Glasgow faculty, Dr. William

Thomson and Dr. Nichol, volunteered advice that was transmitted to William. Natural philosophy was required of medical students at Glasgow, and the physician, Thomson, wanted William to pay "*the greatest attention*" to lectures in experimental physics, to write out lectures and discourses in "the plainest and most attractive terms," and to improve his elocution. People at Glasgow, Dr. William Thomson cautioned, knew that William was an excellent mathematician, but they felt he was too deep to be "popular." William's father urged him to do "all in [his] power to obviate this impression." Dr. Nichol was just as anxious that William devote his attention to everything experimental. He too urged writing out lectures and discourses in "plain, familiar, and popular style," in order to counteract the belief that William was too deep for "the generality of students."[21]

Early in March 1845, Thomson decided on Victor Regnault's laboratory as the best place to learn about experimenting. Regnault, chemist and physicist, was professor of physics at the Collège de France. He had studied at the Ecole polytechnique and the Ecole des mines where he began experimental chemical work. Afterward, he was appointed professor of chemistry at Lyons. His early reputation was made in organic chemistry and, in 1840, he was appointed to the chair of chemistry at the Ecole polytechnique as Gay-Lussac's successor; in the same year, he was elected to the chemical section of the Académie des Sciences. In 1841, Regnault was appointed professor of physics at the Collège de France where he succeeded his teacher, P. L. Dulong. From this time on, he devoted his efforts to experimental physics. He worked on specific heats of a number of substances and on the expansion of gases, and was particularly interested in finding a means to increase the efficiency of steam engines.[22]

When Thomson met Henry Victor Regnault in the spring of 1845, Regnault was thirty-five years old and Thomson twenty. After a tour of the laboratory, Thomson knew he wanted to work there if it were at all possible. The fine array of equipment visible to the visitor had been purchased with government funds, which were provided for apparatus to be used in popular lectures. The government was supporting Regnault's research which was expected to supply data for improving the efficiency of steam engines. Thomson had reason to envy Regnault's equipment, which contrasted with the paltry and well-worn apparatus used for lecture demonstrations at Glasgow. Funds for Regnault's laboratory were cut off for a time after the 1848 revolution, and Thomson was then to realize the disadvantages of governmental support.

After the tour, Thomson went to dinner with one of the assistants

Margaret Thomson.
Photograph circa 1858.

William Thomson.
Daguerrotype, 1854.

Professor James Thomson, Sr.

Sir William Thomson, 1870.

Professor James Thomson,
Lord Kelvin's brother.

James Clerk Maxwell.

St. Peter's College, Cambridge.

Quadrangle of the Old Glasgow College.

Lord Kelvin's lecture room, University of Glasgow.

Portable atmospheric electrometer
invented by Kelvin.

Kelvin's atmospheric electrometer.

H.M.S. *Agamemnon,* Doulus Bay, 5 August 1858,
completing the laying of the first Atlantic cable.
After a painting by Henry Clifford.

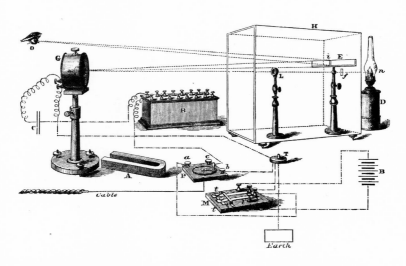

Telegraph submarine cable with Thomson mirror galvanometer.

George Gabriel Stokes.

Peter Guthrie Tait.

Hermann von Helmholtz.

James Prescott Joule.

New University buildings, Gilmore Hill, Glasgow.

Lord Kelvin and his compass.

The *Lalla Rookh*.

Netherhall, Largs.

The *Great Eastern* picking up the Atlantic cable, 1865.

The paying-out machinery of the *Great Eastern*.

Testing the Atlantic cable on board the *Great Eastern*.
Painting by Robert Dudley, owned by the Metropolitan Museum of Art.

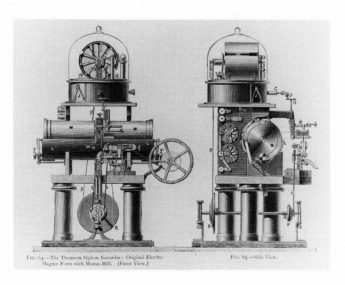

Thomson Siphon Recorder.

Lord and Lady Kelvin, 1906.

Lord Kelvin on a footbridge,
probably at Niagara Falls.

William, James, and Elizabeth Thomson, 1885.
From a drawing by Agnes Gardner King.

and was told that Regnault did not want students as assistants; unlike the chemical laboratories in France, the physicists hired full-time assistants. Thomson worried that his inexperience was a bar to gaining entrance; if that were the case, his second choice would be a chemical laboratory. He became a casual visitor to Regnault's laboratory and soon was given odd jobs such as holding a tube while Regnault sealed it or working an air pump. His father urged him to continue going to the laboratory even if it were only to serve as an extra pair of hands. A letter from Regnault, he wrote, "with reference to *practical* matters might serve you much."[23]

Thomson became one of Regnault's laboratory assistants and was soon hard at work, coming in at 8:00 A.M. and seldom leaving before 5:00 P.M., doing everything from graduating tubes for experiments to helping Regnault with formulas used in calculating results of the experiments. The subject of research was measuring the latent heat of steam under high pressures; the results were of direct use in the design of steam engines. When there was free time, Thomson read books and articles on elementary physics in Regnault's library.[24]

Regnault later supplied the testimonial that Thomson needed as a candidate for the natural philosophy chair, but what he gained from the experience far exceeded in worth the testimonial letter. Already familiar with Fourier's mathematical theory of heat, Thomson was introduced to the new science of thermodynamics, the study of the convertibility between heat and mechanical work. Thermodynamics was to be an area to which Thomson made fundamental contributions. Thermodynamics and electricity were the two major areas of his scientific research. It was in thinking about thermodynamics and electricity in conjunction with each other that Thomson developed his dynamical theory of energy—a major theme in nineteenth-century science. In Thomson's first articles on thermodynamics, he used Regnault's data as demonstration of his mathematical theory.

Regnault's experimental research was as different from lecture demonstrations as the problems in physics textbooks are from original mathematical research. Demonstrations are expository and as such are illustrations for a lecture. They must be visible and simple representations of an idea. Experimental research can be as intricate as the mind of the experimenter can devise. It is up to him to interpret the results. Experiments are an artificial means of exploring nature or searching the unknown. Lecture demonstrations are a visual means of explaining the known. Ironically, the professors at Glasgow may have thought that Thomson was gaining invaluable experience needed to make him a successful lecturer. He never became a successful, popular lecturer. Neither did Thomson become a successful experimentalist. He did not have the talent, and the stay at Regnault's laboratory was much too brief for him to learn the intricacies of that kind of research. He did learn enough about the experimental method to be able to suggest experiments for others to do when there were data he needed. The ability to imagine physical phenomena was expanded by a short but intense apprenticeship with one of Europe's best experimenters. His father hoped that the expense of the Paris trip would make Thomson a better teacher; in Regnault's laboratory, he learned to be a better scientist.

Regnault enjoyed talking to the politically naive Britisher. When Thomson claimed that Britain, even with its aristocracy and rigid class structure, had more liberty than post-revolutionary France, Regnault brought up the matter of test oaths which prevented all but those professing Church of England or Scotland from teaching in the universities. The test oath might be a small matter in light of the fact that there were so many independent positions in Britain, whereas in France, it seemed to Thomson, all scientific posts were controlled by the government.[25] The posts that Thomson had in mind were in British grammar schools, where elementary science was taught and the teacher might, if he were self-disciplined enough, do research in his spare time. In France, the government had bound science with engineering, and teachers had an opportunity to teach more advanced scientific topics than most of the teachers in Britain. Where was scientific research apt to benefit most? In France, where state control and political allegiance tied scientists more closely with technology, or in Britain where the test oath, meaning a form of religious allegiance, restricted university posts to religious conformists? From the middle of the nineteenth century, Britain was in the forefront in advancing scientific theory for reasons more complex than either Regnault or Thomson realized or could have anticipated at the time. Thomson, after all, was to be one of the leaders in the flowering of British science and, as has been shown and will be discussed later, the reasons for his achievements go well beyond the simple matter of test oaths and governmental support.

The second part of Thomson's mission to Paris, making the acquaintance of leading mathematicians, also went well. He met Joseph Liouville, Augustin Cauchy, Michel Chasles, and Jacques Charles François Sturm. After his second visit to Liouville, Thomson decided that he would present him with a copy of his father's *Algebra*.[26] Testimonials from these men would be, according to his father, very valuable in impressing the faculty at Glasgow, but for William, it was an opportunity to talk to mathematicians whose interests were similar to his own. He was received with cordiality granted to a colleague, thus giving him an opportunity to discuss their work in progress. William had learned a great deal from reading the mathematicians' papers, and he now had the good fortune to be able to hear about the thinking behind their original work. The discussions were of the kind he often had with Smith, Ellis, Gregory, and others of the Cambridge school. Conversations with mathematicians of different backgrounds from the British school gave Thomson some new insights, and they in turn learned from the young man who was just beginning to make his reputation.

A few weeks after his arrival, Thomson delivered Forbes's letter of introduction to Cauchy on a Friday, the only day Cauchy received visitors. Cauchy began the conversation by asking Thomson how much mathematics he knew and in particular whether he knew differential calculus, perhaps indicating the Frenchman's impression of the backward state of British mathematics. When Cauchy was assured that Thomson knew enough to understand, he launched into a description of his latest work. Three weeks later, William informed his father of his frequent visits to Cauchy and mentioned that during the visits Cauchy "tells me what he is working at, and all the fine things he is discovering." Cauchy's enterprise was awesome. He had one or two memoirs for weekly meetings of the Institut de France in Paris.[27]

Cayley, at Cambridge, had given Thomson an article to deliver to Joseph Liouville, editor of *Journal de mathématiques*. Thomson was known to Liouville as the author of an article published in his journal in 1844. On their first meeting, Liouville "began directly to work with pen and paper at various subjects of conversation and when I went away," William delightedly reported to his father, "he invited me to return . . . 'pour causer de toutes ces choses.' "[28] Liouville introduced Thomson to Michel Chasles on one of his visits. The name was very familiar to Thomson, for Chasles was one of the mathematicians who had anticipated him in his article in the *Cambridge Mathematical Journal*.

The French mathematicians whom Thomson met were strikingly similar in background; this commonality accounted for their taking the same view on most subjects. The common base was engineering. Cauchy studied at both the Ecole polytechnique and the Ecole des ponts et chaussées. He served three years as an engineer on the construction of the harbor at Cherbourg. During that time, he published some articles on geometry, and with the encouragement of

Laplace and Lagrange, Cauchy gave up engineering and concentrated on mathematics. In 1815, he was appointed to the Ecole polytechnique and, in 1821, began publishing a series of his lectures on analysis and infinitesimal calculus. Cauchy's work revolutionized the teaching of analysis on the Continent and in Britain.[29] His work along with others was the reason why France at the time led Britain in the study of the new mathematics.

Chasles was interested in applied mathematics. He did research in analysis and was also interested in the application of geometry to practical problems in engineering and architecture. Sturm was a professor at Ecole polytechnique where he lectured on analysis and mechanics. The British school of mathematics was centered in Cambridge University where most British scholars came to study pure mathematics, as Thomson had done, whereas the French mathematicians came from different colleges around the country, usually with some connection with engineering and were now centered in Paris. Most of them taught at the Ecole polytechnique. No one from the British school would have been as much at home with the French mathematicians as Thomson, or had such good rapport.

What did they talk about? The main topic, often the source of agitation, was a short memoir by George Green titled *An Essay on the Application of Mathematical Analysis to the Theories of Electricity and Magnetism* (1828). Thomson had seen a reference to the work at Cambridge and, just before leaving, Hopkins gave him two copies of Green's work. Consequently, in Thomson's hands, the work became a despoiler of French complacency. Thomson gave one of his copies of Green's essay to Liouville; it was well placed to stir French thought. Chasles was shown the parts of the memoir dealing with general theorems of attraction, and he acknowledged that Green had anticipated him. Chasles "seemed to be very much struck," Thomson reported.[30]

Sturm heard about the essay and one evening at 10 o'clock, Blackburn and Thomson were astonished to have Sturm pay them a visit. He talked about mathematics for a while but quickly turned the conversation to the essay and, when shown a copy of the work, he looked it over avidly. Thomson pointed out a part in the essay that dealt with a topic that Sturm had written about in 1844 in an article in Liouville's *Journal.* Sturm recognized at once that he too had been anticipated by Green, and exclaimed, "Ah mon Dieu, oui."[31]

An odd drama was taking place that spring of 1845 in Paris. A seventeen-year-old treatise written by a self-taught English mathematician was discomfiting mathematicians who were in the forefront of original research. Green's background was singular and in no way could he be associated with the British or French school of mathematics. This was yet another case in the history of science where a rank outsider had forestalled the scientific community. What is the explanation for Green's apparent prescience?

In many ways, the acquisition of Green's memoir was reminiscent of Thomson's introduction to Fourier. Both Fourier's and Green's theories became basic to Thomson's approach to physical subjects. Both theories were little known in part of the scientific community: Fourier was hardly known in Britain and misunderstood by people like Kelland. Green was unknown in France, and although his approach was different, he had anticipated some important parts of the latest French mathematics. Through his easy assimilation of Fourier and Green, Thomson was able to synthesize two separate ways of thinking. His response to Green was the same as his reaction to Fourier; it was immediate. The first entry in his Paris diary after Hopkins had given him a copy of Green's book was, "I have been studying Green's memoirs on attraction and am just beginning to see through them." What a strange feeling to read a man's work and find that one has "most unwittingly trodden almost exactly in his steps as far as regards electricity." As Thomson studied Green, he found that the work contained all the general theorems on attraction except possibly one. William found some comfort in noting that Green had not dealt with heat.[32]

Conscious of the historical aspect of his subject, Green wrote in his preface that little had been done with the mathematical treatment of electricity between Cavendish's work in 1771 and Poisson's in 1812, "except immediate deductions from known formulae."[33] By this he meant that electric force treated as an effect was measured and quantified, and the relation between these quantities was stated in formulas. The formulas generalized the effect of electric force so that if the quantities of electric charge were changed, or the distances changed, the effect of the force could be calculated. Those were immediate and limited generalizations since the effects being considered were limited. An electric charge either attracted or repelled and the force was calculated by merely multiplying and dividing quantities. This type of quantification amounts to stating experimental results in other terms. Green wanted to treat force as a cause and to describe mathematically the effects. In this way, his approach was similar to Fourier, who treated heat not as a substance but as a cause, and his mathematics described the multiplicity of effects.

Green's object was to determine general mathematical equations to describe what he called "a power of universal agency, like electricity." The other universal agencies, or forces considered as causes, were gravity and heat. In his *Mécanique Céleste,* Laplace generalized Newton's law of gravity and employed the partial differential equation, the use of which, Green declared, was "too marked to escape the notice of any one engaged with the present subject, and naturally served to suggest that this equation might be made subservient to the object I had in view."[34]

Green's object was to give the expression for electrical force in mathematical terms in such a way that this expression of cause, using the logical operations of the differential calculus, enabled the mathematicians to calculate the effects for an unlimited variety of situations. For example, this method was applicable even when the charged electric bodies were of irregular shape. Green's essay introduced the term "potential" as the mathematical expression of the cause of electric force, as Green called it. Potential was calculated by the summation of the masses of infinitesimal particles of a system divided by the distance of the particle from a point. In the case of electricity, the force of attraction exerted on a point by an electrical charge was equal to the sum of the charged infinitesimal particles, each divided by their distance from the point. The potential was the mathematical expression of a cause, just as Newton's definition of gravity was a mathematical expression of a cause and in no way physically defined that cause. Newton said that gravity was a force that was proportional to the product of the masses of two attracting bodies divided by the square of the distance between them. Laplace had shown that taking Newton's mathematical definition as given, the effects of gravity in the solar system could be calculated. Green did for electrical attraction what Newton did for gravity and Fourier for heat. His was another step in the generalization of the idea of force or energy through the use of mathematics. Thomson was to take the lead in the development of the mathematical and physical relation between all forces or forms of energy: thermal, gravitational, electrical, magnetic, and luminescent.

Green extended his concept to the study of magnetic force. Poisson had recently developed equations for magnetic attraction, and his method was analogous to Green's idea of potential where magnetic attraction was expressed in causal terms. Magnetic bodies were thought of as composed of minute particles of magnetic fluid, and Poisson determined the total magnetic force exerted by these particles by using the method of analysis.[35]

Green's work seemed so up to date in the 1840s because the combination of algebra and calculus, known then as analysis, was being used increasingly in the solution of physical problems. In 1828, Green had been a staunch advocate of the method of analysis and wrote that "the applications of analysis to the physical Sciences have the double advantage of manifesting the extraordinary powers of this wonderful instrument of thought, and at the same time of serving to increase them; numberless are the instances of the truth of this assertion." The example that Green gave of the use of analysis as a method of scientific research was Fourier's work. Not only did Fourier discover the general equations for the motion of heat, but these equations were used by Cauchy and Poisson, he said, to give the equation of motion of waves in fluids. Fourier's equations were useful in solving many other problems that Thomson thought were too numerous to be discussed in the article.[36]

Mathematical treatment, as conceived by Green, Thomson, and others of this new school of thought, entailed writing the equations that described the phenomena by using the device of hypothetical fluids or by using analogy. The hypothetical fluids were not to be confused with the imponderable ones of the eighteenth century. The new hypothetical fluids were mathematically describable causes, and the imponderables, although quantifiable, were not mathematically linked to their effects. The equations being developed in the nineteenth century, for example Fourier's equations on heat, were almost universally applicable to the phenomena and held regardless of the shape, composition, or surrounding medium of the bodies under consideration. The equations were so general that they could be used to describe any specific condition by the correct evaluation of the constants of the equation. The distinction was between merely quantifying single effects and using the mathematics of analysis to trace change and, therefore, deal with the course of events. The idea of energy as a cause of a course of events was to be called by Thomson and others "the dynamical approach."

Experimentalists did not accept the mathematicians' view of electrical attraction, branding it too abstract and far removed from physical reality. The mathematicians, for their part, found that the experimentalists' interpretation of the experimental data was not compatible with the general mathematical theories, and they suspected that the experimentalists had allowed concentration on the specific to cloud their search for general theories.

The controversy between the experimentalists and the mathematicians was the subject of a four-and-a-half-hour discussion Thomson had with Liouville. Thomson was asked to write a paper for the

Institut de France explaining Faraday's latest experiments on static electricity.[37] The contrast in the ways electrical force was imagined by way of mathematical theory and by way of Faraday's experimental view made it impossible for one person to hold both views at once. Thomson saw the two views as compatible and what is more, complementary. That was his creative synthesis, and at the time he seemed not to be aware of its importance. In a review of Thomson's *Reprint on Electrostatics and Magnetism,* Maxwell wrote that heat and electricity were so unlike in their effects that no one imagined an analogy between them.[38] But the synthesis of the two points of view, the experimental and the mathematical, was even more unlikely. What Fourier did for the theory of heat had been done for the theory of electricity. Unable to determine the physical cause of electrical attraction, mathematicians instead took the properties of bodies that determined the magnitude of the force. The quantification and mathematical relations between these properties resulted in equations from which the effects of electrical attraction were calculated. The calculations were confirmed by experiment, and the mathematical relations, or laws, were established.

When doing his experiments on electrical attraction, Faraday required a physical conception of the force before he was able to determine laws governing the force. His invention of lines of force acting between bodies that attracted or repelled each other satisfied his imagination and enabled him to draw up his own set of laws. Basic to his view was the composition of the matter filling the space between two bodies under the influence of electrical force. The mathematicians' laws had no such intervening medium. One said that the matter in the space between two bodies was necessary to the exertion of the force, and the other said that whether the space between the two bodies did or did not have matter in it was of no consequence. How could both views be right?

Neither the experimental nor mathematical view revealed the ultimate or physical nature of electrical force. Both had to deal with some form of abstraction. Both anticipated that there was such a cause, and even though the cause in itself could not be described, the laws governing the cause could be determined. In effect, by operating on the premise that there was an answer, scientists could be satisfied with something less than the ultimate explanation. The infinitesimal calculus was a metaphor for this mode of thinking. Although the infinitely large or infinitely small cannot be realized or calculated, the system is based on a belief in the existence of the infinite. As Thomson expressed it later in his life, "I say *finitude* is incomprehensible, the infinite in the universe *is* comprehensible."[39] He said he could not imagine an end to matter or an end to space. In the calculus, the infinitely small unit of matter or of time are components of the mathematical expressions for such things as the movement of heat or the force exerted by an electrical charge. The infinite series of numbers is an abstraction used to describe reality. For Thomson, who had learned the ways of the infinitesimal calculus from his father while he was so young, the use of the symbolism of the infinite was as familiar as the symbolic meaning of "queen" or "Scotland" was to his contemporaries.

Thomson's transferal of this imaginative mode to understanding Faraday was a creative act. Faraday in his own way was abstracting nature in the laboratory in a manner similar to the mathematician. Believing also that simple laws governed the behavior of the unseen reality, the exertion of electrical force, his laboratory experiments were designed to manipulate the unseen, that is, the imperceptible cause of that force. His idea of lines of force was similar to the infinite series used by the mathematicians. Both were abstractions for conceptualizing the real—the finite. Thomson first and Maxwell later conceived of both ideas as abstract and were able to use them in developing theories.

Faraday believed that his results refuted the mathematical theory, which assumed that electric force, like gravity, was transmitted across space. Faraday denied the possibility of instantaneous transmission of force that he called "action-at-a-distance" and insisted that force could act only through the medium of a material substance. Force was an immaterial agency that could produce an effect only by acting on a material substance. The controversy was the scientific equivalent of a long standing controversy in philosophy between materialism and idealism.

So seriously did the French scientists take Faraday's objections that they decided not to make electrical theory the subject of the mathematical prize offered by the Institut de France in 1845. Liouville told Thomson that Poisson, before he died, had wished Liouville to do something to forward the study of electricity by removing the obstacle raised by the controversy. Thomson assured him there was no reason for differences; the disagreement between the two factions was caused by each side not understanding the other's approach.[40]

Thomson's article for Liouville's *Journal* was published later in the *Cambridge and Dublin Mathematical Journal* and was titled "On the Mathematical Theory of Electricity in Equilibrium." "On the Elementary Laws of Statical Electricity" was the title of part one. Even-

tually, there were three parts that together constituted the fulfill-
ment of William's plan that he had made in September 1844. At that
time, he stated that he wanted to write a treatise on the mathematical
theory of electricity. The draft of the paper's introduction that he
wrote in his notebook said:

> The object of the mathematical theory of statical electricity is to deter-
> mine the manner in which electrical intensity is distributed over a con-
> ducting body bounded by a surface of any form, or over a series of such
> bodies in the neighbourhood of one another, and acted upon by any
> quantity of electricity, distributed according to a given law, on non-
> conducting matter.[41]

Thomson began his paper by stating that he intended to show
that the investigations of Britain's two leading experimentalists,
Faraday and William Snow Harris, did not overturn the mathematical theory, but confirmed it. It was impossible, Thomson insisted,
for Faraday's results to be incompatible with Coulomb's experiments; "either method of viewing the subject, when carried sufficiently far, may be made the foundation of a mathematical theory
which would lead to the elementary principles of the other as consequences. This theory would accordingly be the expression of the
ultimate law of the phenomena, independently of any physical
hypothesis. . . . "[42]

In the article, Thomson was speaking as a mathematical theorist
who realized that he could not reject the results of experiment. His
search was for a mathematically idealized representation of electric
force free "of any physical hypothesis." The mathematical theory,
the only true theory, must be free of physical conceptualization, but
the temptation was difficult to resist, as Thomson admitted:

> Now the laws of motion for heat which Fourier lays down in his *Théorie
> Analytique de la Chaleur,* are of that simple elementary kind which consti-
> tute a mathematical theory properly so called; and therefore, when we
> find corresponding laws to be true for the phenomena presented by elec-
> trified bodies, we may make them the foundation of the mathematical
> theory of electricity: and this may be done if we consider them merely as
> actual truths, without adopting any physical hypothesis, . . .

And here his imagination would not be stilled

> although the idea they naturally suggest is that of the propagation of
> some effect by means of the mutual action of contiguous particles; just
> as Coulomb, although his laws naturally suggest the idea of material
> particles attracting or repelling one another at a distance, most carefully
> avoids making this a *physical hypothesis,* and confines himself to the con-

sideration of the mechanical effects which he observes and their necessary consequences.[43]

Thomson examined the results of Snow Harris's experiments and
showed that most of the results agreed with the mathematical
theory—if properly interpreted. Where Snow Harris believed there
were differences, Thomson insisted that he had not taken proper
precautions in doing his experiments. Whether, as Thomson claimed,
it was a necessary precaution to note that "the distance between re-
pelling bodies must be considerable with reference to their linear
dimensions, so that the distribution of electricity on each may be
uninfluenced by the presence of the other" depended on one's view
of the important factors in electrical attraction. Experiments are in-
terrogations of nature, and they are designed according to the questions the experimenter wants to ask. What Thomson said, in effect,
was that had Snow Harris asked the right questions, his conclusions
would have agreed with the mathematical theory. That was the crux
of the controversy. What was the real nature of electricity?[44] Was it
best described as a mathematically describable cause or by means of
the effects it produced?

Snow Harris drew conclusions from his experiments that he saw as
direct contradictions of Coulomb's theory. To Snow Harris, it seemed
that electricity was too various in its effects, and a universal law such
as Coulomb suggested did not hold. Electricity, Snow Harris insisted,
was "an extremely intricate phenomenon" and he implied that this
agency, indescribable by generalizations, was bound to elude mathematical analysis. No gains were made by the mathematicians' approach: instead their work gave "rise to equations [that were] extremely complicated—in some cases very impracticable. . . ."[45]

Thomson, the theoretician, believed that mathematical analysis exceeded all attempts by experiment to understand and explain the
phenomena. Experimental data were the starting point, the contact
with reality, but the mathematical imagination was the only way to
arrive at general theories that were finally independent of any physical hypothesis. The theory was constantly tested and challenged by
experiment. Abstraction free of reliance on physical hypotheses was
the ultimate explanation toward which science progressed.

The trip to Paris was necessary to give Thomson a European reputation, but more in the sense that Hopkins meant than in the sense
that his father and the Glasgow faculty had in mind. To that center
of mathematical studies he brought his copy of Green's essay. As a
result, Thomson and the essay became the focus of a lively discus-

sion. The request by Liouville for the article enabled Thomson to make, for the first time, an explicit statement of the metaphysical basis for his science. For the training in experimental work that he wanted, he found Regnault's laboratory, where practical work in thermodynamics was in progress. William went to Paris specifically to gain practical experience, but circumstances made the trip doubly useful for continuing his scientific research. The discussion over Green's essay helped Thomson to see the relation between the physical and mathematical theories of electricity. Through his service in Regnault's laboratory, he gained a physical view of heat, thermodynamics, to complement Fourier's mathematical view. These ideas and the dualistic approach to theory were the foundation for his life's work.

CHAPTER FIVE

Teaching and Research

William Thomson was reveling in the beautiful days of his last two weeks in Paris. "This is a glorious morning, and so I have no time to lose, as I am going out for a long walk, and intend to bathe in the Marne." In a hurried scrawl, he also informed his father that he would have to be in Cambridge no later than 8 May in order to help Frederick Fuller with the Peterhouse examinations. William hoped to have one or two pupils who would begin studying during the early part of June. He would then have enough time for his work and be free from the beginning of September to 26 October.[1]

Dr. Thomson wanted to hear from his son when he reached London and again when he reached Cambridge. "When in London," he said, "call ... Mr. Ogilby and get him to introduce you to some leading scientific men there such as Faraday if he can, other members of Royal Society, &c., &c., as you aught to be acquainted with English as well as French savants."[2] There is no record of William following these instructions.

William Thomson postponed a reunion with his family until the end of the summer, but he did look forward to a Cambridge reunion of sorts in June. The 1845 meeting of the British Association for the Advancement of Science was to be held there. Dr. Thomson and William Bottomley, Anna's husband, were planning to come, and Dr. Thomson planned to ask other members of the Glasgow faculty. He advised William to urge his Parisian friends to come. "James must come," William insisted, "the place seems to agree with him very well, and it will be a pleasant trip for him. . . . If he is inclined for a vacation from his work, he can remain with me as long as he pleases." William promised to get accommodations at Peterhouse for his family as well as for any others that Dr. Thomson brought, that is, except Anna, a mere woman, who was "not worthy of such an honour." But he offered to get her "a room in Fitzwilliam Street."[3]

James could not come to Cambridge. John wrote to William that James "has to avoid animal food, and exercise, and everything in fact that could excite the circulation; as it is by reducing this that a cure may be effected. As of course all your letters are read by all of us; the less you say as to what you may think of James's health and prospects the better; for it cannot but put him out of spirits to hear such things as you said in your letter."[4]

Since father was still paying the bills, he wanted to know what William's expenses were likely to be in Cambridge, "also whether there is any probability of a fellowship becoming vacant soon, and whether you are likely to get any pupils."[5]

Dr. Thomson need not have worried, for no sooner had William arrived in Cambridge when he received his Smith's prize money: £24.5.5 net. He had four full-time pupils and one half-time[6] and was elected a fellow of Peterhouse, contrary to his expectations. He was elected on 28 June, four days after his twenty-first birthday. Dr. Thomson had just left Cambridge, where he attended the meeting of the British Association, and William wrote to him, "I now wish very much that you had remained, but I did not like to press it before on account of the great uncertainty. . . . You must know very well how glad I am of it." The fellowship was worth about £200 a year plus rooms in college.[7] William's father was elated by the news:

> Long may you live to enjoy your appointment and other good things to which I trust it will be a prelude and a stepping stone. What reason you have to be thankful that you have had health, abilities, and opportunities to enable you to get forward so far at so early an age! At your age I was teaching eight hours a day . . . and during extra hours—often fagged and comparatively listless—I was reading Greek and Latin to prepare me for entering College, which I did not do till nearly two years after.[8]

The fellowship was a high step and the highest achievement for some who passed through Cambridge. However, a fellowship with its benefits was a misfortune for some who settled into indolence. Thomson's brother-in-law, William Bottomley, was certain that "to [William] I am sure a Fellowship yields no such temptation. . . . " Bottomley thought that with money worries removed Thomson would concentrate on his research. At the end of his probationary year, the bishop of Ely installed Thomson into his fellowship and warned him not to be like so many others who frittered away their time. The bishop urged Thomson to continue his studies and original research in science.[9]

The news of the fellowship was a welcome surprise to the family.

Although James knew that William intended to apply for a fellowship on his return from Paris, "[he] thought [it] would be given as a matter of course to one of [William's] seniors."[10] Robert wanted to know whether William dined at the fellows' table and looked down with "sovereign contempt" on his inferiors. Robert wrote that Anna was "in a state of great anxiety that you should get a beard fast so as to make you more imposing." And in celebration of the fellowship, "Papa has at last agreed to take Punch. . . . " Elizabeth, now a minister's wife, struck a dour note: "But in all your success you must not forget that life is transitory and only a time of trial and apprenticeship, as I have some where seen it expressed, to fit us for eternity."[11]

At the 1845 meeting of the British Association, Thomson read his paper titled "On the Elementary Laws of Statical Electricity." Afterward Faraday raised some questions about the paper with Thomson.[12] Dr. Thomson thought the paper "went off very well," but the meeting as a whole must have been a disappointment, at least according to William's brother John, who wrote that he "was sorry to hear of the poorness of the meeting. I fear the glory of the Association is gone, or at least will soon be gone. Unscientific people seem to be tired of it, and also many scientific people. . . . "[13]

A mark of Thomson's growing status in the scientific community was his selection as editor of the *Cambridge Mathematical Journal* to succeed Robert Leslie Ellis who had assumed the editorship because of Gregory's illness. Founded in 1837, the *Journal* had successfully fulfilled Gregory's hopes that Cambridge, a center for mathematical studies, would be the locus for a publication that encouraged original research in mathematics.[14] The *Philosophical Magazine* and the *Transactions of the Royal Society* were devoted more to physical subjects. Thomson had plans for broadening the *Cambridge Mathematical Journal* so that it would resemble Liouville's *Journal*.[15] Thomson's first action as editor was to include Dublin in the name of the journal. His choice was astute. There had developed in and close to the University of Dublin a group of mathematicians who were influenced by the French school. Richard Townsend and George Salmon were followers of Michel Chasles, whom Thomson had met in Paris. The Dublin school also included some highly original thinkers. In the 1840s, W. Rowan Hamilton was in the process of inventing quaternions, and George Boole was developing his new mathematical logic. Although not all these men were connected with Dublin University, their names were associated with the city and with Ireland.[16]

William complained to his Aunt Agnes that in addition to his other work, he was "having a great deal of trouble, and [was writing] an immense quantity of [letters] about setting the Mathematical Journal going on an enlarged scale." One of William's troubles was dealing with rival publishers,[17] a problem he settled by giving the publishing contract to Macmillan. He received advice on how to improve the *Journal:* William Bottomley wrote that he spoke to someone who would resubscribe if he were assured that copies would arrive punctually; De Morgan wrote and complained about the number of printer's errors in the old series.[18]

Thomson solicited advice about the name change. Boole preferred the general title *Mathematical Journal.* So did Archibald Smith, whose opinion was that "if the title is changed [to] show that it is not meant for Cambridge contributers [*sic*] alone I would like to have it so selected as to show it is not meant for *university* contributers alone." Ellis wondered whether a connection with Dublin could be "indicated by getting the name of some Dublin publisher on the title page." Arthur Grant wrote from Dublin: "My own leaning, and I think I may say Sir Wm. Hamilton's, is in favor of the name 'Cambridge and Dublin.' It appears to us most likely to conciliate the support and to stimulate the energies of men connected with this university. And I would venture to urge that the exclusiveness of title can do you little harm. . . . If you decidedly object to 'Cambridge and Dublin' I would suggest 'British and Irish.' . . . "[19]

The first volume of Thomson's *Cambridge and Dublin Mathematical Journal* was published in 1846. It contained articles by Boole, Hamilton, and Townsend in addition to the group from Cambridge, which included Ellis, Cayley, and Thomson. There was also an article by a fellow of Pembroke College, G. G. Stokes, who was beginning to make his mark in science. Liouville contributed an article on static electricity in which he mentioned George Green's work—further evidence of Thomson's growing influence.

Soon after he returned from Paris, Thomson began a new mathematical notebook. The work showed a carry over of William's interest in the subjects he had discussed with the French mathematicians. The notebook began with an attempt to verify formulas suggested by the mathematics that had been used for electrical attraction between charged planes. The entry on 5 July 1845 concerned the distribution of electricity on spheres. The problems were abstract mathematical, and they dealt with abstract geometrical shapes rather than experimental or physical instances of electrical attraction. Gradually he progressed from his former analogy with heat to an analogy with light, first referring to the electrical attraction between charged spheres as being like successive influences or reflections.[20]

A month passed between each entry in the notebook, and William explained the hiatus as being due to spending a good deal of time tutoring students at Cambridge. During that time, however, he had been thinking about the distribution of electricity on spheres and planes and had gotten "an immense addition of light" on the subject. The geometrical conditions he was considering were becoming more complicated. What was the distribution of electricity, and the resulting potential or force, on two spheres intersecting at an angle and subject to an external charge of electricity? The effect was analogous to the kaleidoscope; the solution of that problem enabled him to verify the formulas with which he started in July.[21]

By 17 August, William was "beginning to get quite anxious to get home again" and was "longing to see" the whole family. They were all at Knock that summer. He arrived there the following week and was able to get back to his notebook on 9 September when he carried forward the abstract conception of intersecting spheres subject to the force exerted by an external electrified point. A new analogy with light was used this time, which he called "the principle of images." Force, in geometrical terms, was analogous to the reflected image of a lighted object.[22] There was no physical conception of either the light images or the analogous electrical image.

The principle of electrical images and the kaleidoscope were analogies suggested by the mathematics. Faraday was to announce the physical relation between light and magnetism, and the idea would be seized upon by Thomson. But for a time, he was still under the influence of French mathematics. The distribution of hypothetical electric or magnetic fluids as a way of calculating force used Green's idea of potential and was in that way mathematically analogous to gravity. However, the mathematical equations were abstract, that is, general and independent of any physical interpretation as to the cause of magnetic and electric force.

Laplace's analysis and Poisson's formulas proved helpful in these investigations, and it seemed reasonable to Thomson to send the results of his research in note form to Liouville for publication in his *Journal*. Liouville liked the abstract mathematical idea and included a comment in his *Journal* on Thomson's theory of electrical images. In a personal note at the end of his mathematical development, Liouville spoke of the high esteem he had for Thomson's talent.[23] The mathematical idea of images was of little interest in Britain. Thomson's note and Liouville's additions were neither translated nor reprinted in any of the British journals. A one-page nonmathematical description of the idea of images appeared in the *Report of the British*

Association for 1847. Although it was little used, the idea was much appreciated.[24]

When William Thomson returned to Cambridge during the early part of October, he received word that must have made him wonder if he was getting absentminded at an early age. He had taken one of his brother John's shoes instead of one of his own; now he had an odd pair of shoes. But he did remember to inquire for his father as to which college would be the best for one of Dr. Thomson's good students. The question was which college would have a fellowship vacancy when the student finished his degree. William wrote to his father that the student would probably get a scholarship at the outset if he applied to Trinity and would be sure of getting a fellowship later on. A disadvantage at Trinity was William Whewell "who meddle[d] with everything in a most mischievous manner."[25] A week later, William wrote to his father:

> The list of new Trinity fellows has appeared today, and is a specimen on the whole of Whewell's usual discrimination. Neither Blackburn or Grant has got a fellowship I am sorry to say.
>
> The two mathematical papers, one of them set by Whewell, and another by a very weak man, have been of the mildest description. . . . Hopkins says he never saw such "lean papers," and most certainly it would be impossible to . . . give a worse *style* of examination for men who have had a year's reading after the senatehouse. Every one says that the Trinity fellowship *mathematical* examination was some time ago of a very high and philosophical kind, but it has now got to a very low ebb. . . . The junior fellows of Trinity are beginning to feel strongly the necessity of keeping up the character of their body, and the senior fellows, in whose hands the fellowship election is placed, will probably begin to feel a little in the same way, and even the despotic Whewell may have to succumb to public opinion, and let proper men set the mathematical papers.[26]

Dr. Thomson's student went to Peterhouse.[27]

Thomson was content to stay in Cambridge that year and did not want to leave unless it was absolutely necessary.[28] He had as many private pupils as he wanted and still had two days a week free. The college lectures which he gave every morning at eight pleased him more than the private lessons. It was not until 31 October that William found time for his favorite relaxation, music; he was able to play one of the cornopean parts in the Cambridge University Musical Society Orchestra. During his stay in Paris, he had relaxed in a similar way. He had spent his Easter vacation taking cornopean lessons. Dr. Thomson was annoyed with William's penchant for relaxing with music and thought that he should spend his time more profitably. William justified this expenditure of time and money by telling his father that he had "the benefit of a tolerable lesson in French," since the music teacher spoke no English.[29]

At Cambridge in 1846, Thomson earned more than enough money for his needs; consequently, he only had to justify what he did to himself. A friend of his in the musical society thought that Thomson was "most indefatigable" in his musical studies. Thomson was now playing the French horn as well as the cornopean.[30] During a week in London in 1847, Thomson took several French horn lessons and bought a horn. He also attended a rehearsal and concert of the Philharmonic Society with some Cambridge friends. And after hearing Jenny Lind in concert, he wrote to his brother James that he "was as much delighted as possible in every way with her."[31]

In a notebook entry of July 1846, Thomson summarized his work since the previous July and expressed discontent with the progress of his research. In addition to his published correspondence with Liouville, he had written a number of articles for the *Cambridge and Dublin Mathematical Journal*. In his notebook, some of his research was given up as illusory and other attempts at solutions were abortive. William was becoming anxious to see results of his work published at once, "seeing that I have made many abortive attempts to commence my *treatise* on electricity, which I have intended to publish in parts in the journal, and the prospect of getting it ready is now rather more distant." During the 1845–1846 college year, he lectured third-year students at Peterhouse for one hour a day, "effectively a good deal more," he complained.[32]

Demands on his time by other matters explain in part why William was making unsatisfactory progress with his research. The other reason was that he was not content with the abstract mathematical approach to electric attraction; he remarked in his notebook in the summer of 1846 that since that spring he had been attempting to connect the propagation of magnetic and electric force with what he called "solid transmission of force." By the term "solid transmission" he meant an analogy with the physical or mechanical exertion of a force.[33] The impression that the French mathematicians had made on him was rapidly receding; Faraday's latest experimental research contributed to the restoration of balance between the mathematical and the physical in Thomson's work.

In December 1845, Archibald Smith wrote to William Thomson and asked him what he thought of Michael Faraday's latest discovery, the direct relation of electricity and magnetism to light, which was reported in a letter to the *Times*[34] a day after Faraday announced his discovery to the members of the Royal Institution, where he was Fullerian Professor of Chemistry.

Faraday's discovery of the physical relation between light and magnetism came after a long period of experimenting. His method, as in the case of his discovery of electromagnetic induction, involved the dogged persistence in trial after trial of using many different materials under as many different conditions as he could think of, in pursuit of the effect that he knew existed. He was certain that the intervening medium affected the transmission of magnetic force and that a transparent medium should have a special effect. On 13 September 1845, he found that magnetism affected polarized light in a transparent medium, and his notebook entry for that day claimed: "Thus magnetic force and light were proved to have relations to each other. This fact will most likely prove exceedingly fertile, and of great value in the investigation of the conditions of natural force."[35] Faraday's "exceedingly fertile" discovery forged another link in the attempt of nineteenth-century scientists to relate all forms of "force," or in present-day terms, all forms of energy.

On Saturday, 24 January 1846, the day after Faraday gave an illustration of his recent research in magnetism and light at a Friday evening lecture at the Royal Institution, William Thomson came to London and had a long conversation with Faraday "in which he explained to [Thomson] some of his recent discoveries in magnetism."[36] In April, Thomson wrote to his brother James and urged him to read about Faraday's latest discoveries. Accounts were to be found in the *Athenaeum* or newspapers and in the *Philosophical Transactions of the Royal Society*. The experiments had excited considerable interest, and a report of Pouillet's confirmation of the results appeared in *Comptes Rendus* a few weeks after Faraday's results were printed. After reading a reprint that Faraday had sent to Cambridge, Thomson wanted a more detailed and direct account of the research. He visited Faraday at the Royal Institution and had a long conversation with the experimenter.[37]

Faraday had no way of knowing how far or in what direction his discovery would take scientific investigation. Thomson's unique interpretation of the results made Faraday's discovery the opening wedge to a new way of understanding the relation between all forces: luminescent, magnetic, electric, mechanical, and gravitational. At the very moment of Faraday's success, Thomson was following his own thought along paths determined by his conviction that in these forces as in all nature there was unity. The discovery of this unity, Thomson believed, was the goal of science. The effect that Faraday discovered was important to nineteenth-century scientists and Thomson in particular because it was an important clue to the unification of science by showing that all energy was related.

Settling at Cambridge, Thomson was thinking of remaining there for two or three years—living and working in the manner he had established. His future was not yet determined. He had written to a friend that he thought natural philosophy was "much more agreable [*sic*]" to teach than pure mathematics.[38] William had prepared for a teaching career, but the chair at the University of Glasgow was always in the background.

In February 1846, Dr. Thomson wrote to William about a teaching vacancy in Glasgow. The position paid well and was one that a person William's age rarely attained. The post was teaching high-school mathematics, and Dr. Thomson saw that one objection was the status of the situation. The matter was for William to decide, but his father thought that the notion of status carried no weight. Teaching in high school had served as Dr. Thomson's apprenticeship as well as that of several other professors at the University of Glasgow. William, his father wrote, was to deliberate on the matter. At first Thomson wanted to apply for the job. The income was tempting, and he thought the status was all that might be desired. But he decided to stay at Cambridge for the time being, believing that his research would suffer on account of the numerous hours required for teaching at the high school.[39]

The large number of applications for the position indicated that in spite of the teaching load, it was a desirable post; the opportunities for teaching and research in science were very few. The city council of Glasgow was searching for a man who would teach arithmetic, mathematics, and nonmathematical geography in a simple, interesting, and tasteful way. Thomson suggested J.A.L. Airey, second wrangler in 1846 and Hopkins's student, for the job. Airey decided not to become a candidate and went to the Continent instead as a young man's private tutor.[40] For one who stood so high in Cambridge's mathematical honors, Airey did not have an eminent career in science. He took a position as mathematical tutor at a grammar school, and years later as Reverend J.A.L. Airey he became rector of St. Helen's Church in Bishopsgate.[41] As James wrote, the position in Glasgow was finally "thrown away on an inferior man, Brice of Belfast, through the party spirit of some of the electors and on account of the absence in London of some of the more sensible ones."[42]

William Thomson might well have accomplished good things in science while working at the high-school post if George Boole can be taken as an example. Boole, an original mathematical intellect of the nineteenth century, was a master at a boys' school for several years, where he taught more classes in classics than mathematics. In 1849, he was appointed to the mathematics chair at Queen's College in Cork largely through the efforts of Thomson who helped in the testimonial campaign that every candidate had to conduct. *The Laws of Thought,* Boole's most original work, was published five years after his appointment at Queen's College. But his theory, which found increasing application with succeeding generations of mathematicians, was first published in pamphlet form in 1847. Although Boole complained to Thomson that he was discouraged by the burden of work at the boys' school, he managed to find enough free time to develop his ideas there.[43] Might not Thomson have followed the same pathway? What else was available to him? He wanted neither the church nor the bar, and if he taught mathematics his teaching was expected to be "popular" and therefore far removed from his research interests.

The prospects for a career in science were not promising in mid-century Britain, as Charles Babbage described them in his *Reflections on the Decline of the State of Science in England* (1830).

> Let us now look at the prospects of a young man at his entrance into life, who, impelled by an almost irresistible desire to devote himself to the abstruser sciences, or who, confident in the energy of youthful power, feels that the career of science is that in which his mental faculties are most fitted to achieve the reputation for which he pants. What are his prospects?

Very small, according to Babbage's assessment. There were no situations in government and none anywhere else in the society of any account. A university chair was a slim possibility "to which he may at some distant day pretend," but these were not numerous, and although the salary might be enough for a single man it was rarely enough for supporting a family. There was no choice for a young man without independent means. He had to choose the law or another profession for which he had little talent, in comparison to his scientific capabilities. Babbage concluded that the loss to the young man was severe, but to the country it was an even greater loss. "We thus, by a destructive misapplication of talent which our institutions create, exchange a profound philosopher for but a tolerable lawyer."[44]

Notwithstanding Babbage's trenchant criticism, there was no reason to believe that providing more university posts or establishing government positions for scientists would enhance British science. Most of the best work was done outside the universities, and the few men with university posts who were noted by Babbage contributed hardly anything of consequence. It was customary for the college lecturer at Cambridge to use up his time outside lecture hours by taking private students, just as it was customary for the Glasgow professor to conduct lecture series and write textbooks to augment his income instead of doing original research. There was very little in common between undergraduate lectures in physics and work in the newest areas of investigation.

Many teachers preferred doing other than scientific research with their available time. Thomson's friend Fischer, who had been fourth wrangler in 1845 and would be appointed professor of mathematics at St. Andrews, was still coaching students at Cambridge when he wrote to Thomson that "I am afraid you will think me very unprofitably engaged if I tell you that I spend many a leisure hour on Göthe [sic] and German literature [a taste that Fischer had introduced Thomson to during their Cambridge days]; particular [sic] on Aesthetics rather desirous of cultivating my taste for works of art, and of gaining a consciousness of the pleasure I feel in them."[45]

Recognizing that teaching and research were not complementary, William Thomson realized that he must divide his time. The association with science through the teaching of fundamentals was no advantage to one engrossed in new ideas. If Thomson were to be appointed to the chair at Glasgow, as his father hoped, the advantage was a six-month university year. A man who was interested in research had from 1 May until the beginning of November to concentrate on his investigations.

The chair of natural philosophy at the University of Glasgow became vacant when William Meikleham, who had been ill for five years, died. Dr. Thomson wrote to William at Cambridge: "The enclosed notice [of Meikleham's death] must put you into active and energetic motion without delay." The campaign had begun in earnest, and William must get testimonials "from respectable men."[46]

William and his father had been thinking about the job for years. Dr. Thomson had always tried to enhance William's qualifications for the chair—the trip to Paris, having William meet the right people. On his trip to London in January, when he spoke to Faraday, William had two letters of introduction to present: one was to Professor A.W. von Hofmann, the first professor of chemistry in the newly opened Royal College of Chemistry. The college's charge was the advancement of chemistry, especially in its relation to agriculture and industry. Like Glasgow, the emphasis was on practical applications. Professor Hofmann showed Thomson the laboratory, and he stayed to watch some of its operations. William couldn't present the other letter of introduction because he had several dental appointments during his stay in London.[47]

The sudden vacancy, though not unexpected, took William by surprise. Just before receiving the news, William wrote to his father about staying in Cambridge to finish some work which he found difficult to do while teaching. When the articles he was working on were completed, then perhaps he would take a short tour of the Continent. The news of Meikleham's death unsettled his plans and any thoughts he had about staying at Cambridge for two or three more years. Although he assured his father that he was setting out immediately to procure testimonials, there is an evident lack of zeal in William's insistence that he thought "it will not be a good plan to attempt to get too many" testimonials.[48]

In June, William's father wrote to complain of the slowness with which the testimonials were coming in. Was William resting on his oars? No, there were other matters he had to attend to—there were articles to write and students to tutor. William planned to remain in Cambridge up until the time of the election. That would not do, his father retorted; he must arrange to be in Glasgow considerably earlier. After reminding William that he did not even know who the other candidates were, he told his son that someone who had read William's letter said that "it [did] not appear to be the note of a person who [was] trying to obtain a valuable appointment for life, and who [was] perhaps at a principal crisis in his career." William *had* heard reports of other candidates. "All these things, in addition to the general uncertainty of the prospect, have been making me very anxious," he confided to his father. "And you may be quite sure that (although you think I am too cool on the subject) nothing which I can do to increase my chance will be left undone."[49]

Dr. James Thomson may have been feeling that the position was not the best one for his son. When the session ended, Dr. Thomson wrote to William that the year had been the most unpleasant one in his experience at Glasgow. There were riots during the year, caused, he said, by "turbulent lads" who unsettled the minds of many of the students and led them to "idleness, giddiness, and other misconduct in the Classes." William's father found the cure to be severe discipline with the result that his class had "the appearance of a church during the time of service."[50] How would a mild-mannered youth like William Thomson manage these unruly boys?

Dr. Thomson exhorted William to get into "active and energetic motion without delay." He was to think of all possible people for testimonials. What of Faraday, Sabine, Sir W.R. Hamilton, and Snow Harris? Among foreigners, what of Liouville, Cauchy, Lamé, Sturm, Regnault, Gauss, and Chasles? William was to try to obtain testimonials that would obviate the impression that he was "too deep and too rapid" in teaching. Private letters to Glasgow professors would have a greater effect than mere testimonials. The dean and lord rector of the university had the right to vote in the election for the chair. Could William "get [to] them?", his father asked. The dean was a *vain* man" who would be flattered by a letter from a great or learned man. The campaign was under way in earnest, and William was to write home frequently.[51]

Dr. Thomson felt that it would "be greatly . . . regretted if you get nothing from Faraday. . . . " However, William informed his father that Faraday had made it a rule not to write testimonials. "He may however perhaps do something for me, and I know he is very favourable to me (unless very polite)." Nevertheless, Dr. Thomson urged William to get a recommendation from Faraday either directly or indirectly. After calling on Faraday and finding he was out of town, William decided to attend Faraday's Saturday afternoon lecture[52]

> which, though on his new discoveries, was very popular, and very interesting to miscellaneous assemblages of fashionable people. I spoke to Faraday after the lecture, and told him that I am a candidate for the vacant professorship. As I anticipated he makes a rule not to give testimonials (and he gave the same opinion as others about the badness of testimonials in general) but he wished me success.[53]

There was nothing to do but take the indirect approach. William sent his father a letter he had received from Faraday in December 1845, "which may possibly be of some use."[54]

When Dr. Thomson was confined to bed for a few days, his instructions to William were written by James or John. How many testimonials should there be? William thought only a few from people he knew well was best. His father agreed with him but also felt that some members of the faculty were impressed by the number of letters as well as by the importance of the writer. Letters should be sought from Hopkins and Cookson, the moderators and examiners for the Tripos, the Smith's prize examiners, and the master and fellows of Peterhouse. What of Earnshaw, Ellis, Peacock, and Challis? Their letters would be most valuable if they emphasized

William's acquaintance "with the more popular parts of Nat. Phil." and would be of great service if they appeared to be unsolicited. William told his father that he was writing to Boole for a testimonial. "His opinion will *really* be very important as he is a mathematician of a very high order . . . but it is not very likely that he will say anything very strong (though we know one another well) and besides I suppose his name is not known at Glasgow." It was time for William to announce his candidacy, and his father gave examples of the correct way to address the electors. The testimonials must be printed, and he suggested that they be gilt-edged.[55]

William Thomson was well known to the Glasgow faculty. Some professors had taught him; some knew him personally. All were aware of his achievements at Cambridge and of his growing reputation. A proud father and the local newspapers had spread the news. Nevertheless, the election was not assured, according to his father's assessment. Conversations with faculty members who favored William's candidacy revealed a concern that he was unable "to bring down [his] instructing to the capacity of ordinary students." Such was the report that came from Cambridge where William had private students. Dr. Thomson urged, "You *must* take care to cure the evil, if it exists; and if not, to teach so simply, clearly, and slowly, that you may be able to get decidedly good testimonials on that point. *Do attend to this above all things.*" Another professor wanted to hear William read a "*non-mathematical*" paper or give a lecture because he wondered about William's "timidity and want of effective elocution." William was being made uncomfortably aware of his shortcomings. Since this professor was to be in Cambridge soon, Dr. Thomson wanted William to use the occasion to show that he was not shy. "Do *speak out,* to him and others, freely, and without hesitation." If one could not hear a paper by the candidate, a friendly professor asked, perhaps he might supply a composition on a common subject. Dr. Thomson was willing to do almost anything, but this last request amused him. He asked if William had an old logic-class essay around and urged his son to do all in his power to counter the impression that "[his] ideas and expressions are bound up in the icy chains of X's and Y's, +'s and −'s."[56]

Dr. Thomson was a tireless campaigner. He took time to visit his grandson James Thomson Bottomley in Ireland, and even then he kept in touch with William. The matter of the testimonials took most of his attention. He constantly advised William to make the necessary contacts. It was decided that the printing of the testimonials would be done in Glasgow, and Dr. Thomson was to check them. He

visited the faculty, as well as those in the administration who remained in Glasgow that summer. He also sought advice and help in
lining up votes from those who were friendly. The undecided electors were asked not to promise their votes to any other candidate
and to wait until William's testimonials were submitted. Dr. Thomson worried about other candidates, both those who openly can-
vassed and those who were being put forward by their friends. With
all this effort, one can understand his exasperation when William
wrote that he had "been feeling very much a dislike of testimonials,
and want of faith in their power. . . . " William thought the electors,
not the candidates, should write for testimonials in order to obtain
fairer results.[57]

Was Archibald Smith a candidate? The indecision over that question kept the Thomson camp on edge. If Smith decided to stand, he
would be a formidable rival. He was a fellow of Trinity College in
Cambridge. He had been senior wrangler in 1836 and first Smith's
prizeman. After taking his degree, he had helped found the *Cambridge Mathematical Journal* with D.F. Gregory. Smith, like many
other men from Cambridge with mathematical honors, went to the
bar. In 1846, at the age of thirty-three, he was well established in
London as a property lawyer. He continued with his mathematical
interest and was later elected to the Royal Society on the basis of his
work in magnetism. Archibald Smith and William Thomson were
good friends. They had known each other since 1841 when Gregory
brought Smith to meet William who shortly after described the occasion to his father: "It was certainly a great honour for a freshman of
St. Peter's to have two fellows of Trinity calling on him! They staid
[*sic*], I suppose nearly three quarters of an hour in my rooms, looking into my books and talking on various subjects connected with
them." It was upon inquiring for a testimonial that William found
that Smith himself might be a candidate. The situation was awkward,
as Smith recognized.[58]

Smith's father was a wealthy West Indian merchant living in Glasgow, and he campaigned actively for his son. Dr. Thomson reported
to William that "old Mr. Smith" had returned from abroad and was
moving around Glasgow, using every means in his power to forward
his son's candidacy "*without much scrupulousness.*" William now realized how grim the campaign was. Dr. Thomson did not trust the
older Smith and called him "a cunning, wily man." Of William's
friend Archibald, he had a better opinion, "though not an *unqualified*
one."[59]

Smith was uncertain about becoming a candidate for the position at the University of Glasgow and did not even want it known that he was thinking about it, out of concern for his practice. His clients might think that he was not serious in his intentions about the law. Clients did not think, one assumes, that a lawyer was not in earnest when he published mathematical papers. Edward Lushington, one of the electors, told William that he was surprised at Smith's candidacy, having heard that he was a successful lawyer. Smith wrote to Dr. Thomson that he did not want his indecision to injure William's chances, and he knew that he should make up his mind. The professorship, Smith wrote, "has so many attractions for me, attractions which a few years ago would have been irresistible."[60] Nevertheless, Smith neither withdrew nor placed his name in candidacy. As William foresaw, Smith did not have a chance unless he decided to campaign actively.

On 8 July 1846, Dr. James Thomson was in London giving testimony on behalf of a parliamentary bill to remove the test oath in the Scottish universities. He took the occasion to see Archibald Smith and reported to William that Smith was "very silent and distant." "He had not yet come to a decision as to the course he would pursue." One of the Glasgow professors who was also in London for testimony was approached by Dr. Thomson for the first time about William's candidacy. The professor, Dr. Gray, told Dr. Thomson that he thought the chief ground for opposing William's appointment was that he was a professor's son. Here was still another problem, Dr. Thomson realized, that had to be handled by indirect means. Dr. Gray thought that William's credentials were excellent and that his qualifications were the best. Therefore, Gray was convinced that if William did not receive this appointment he "would soon get another."[61]

All the effort put into the four-month campaign bound the Thomson family's attention; world affairs were in second place. No matter what the tensions were at home or abroad, Dr. Thomson's letters to William rarely strayed from talk about the university and the town. Aunt Agnes wrote about friends. She also wrote about the new outbreaks of trouble in India, because they affected Glasgow people, and reported the death of a young man from the city. He was the third Glasgow man who was killed in the fighting in India. But such comments were rare. The whole turmoil of mid-nineteenth-century Britain went without notice in the correspondence between members of the Thomson family. In September, Aunt Agnes was thinking only of the election for the chair of natural philosophy. She would lay awake counting the votes. All but one vote seemed assured for

William, but from an early age Aunt Agnes had found that she had usually been disappointed in the things she wanted most. Still, she wrote, "The thought of what you and your father would suffer if disappointed now is so painful that I cannot bear to contemplate the mere idea. . . . " Even William Bottomley, William's brother-in-law, did not forgo giving advice: "I quite approve of you[r] making every exertion to obtain the election. . . . No efforts indeed can be too great."[62]

The election for the chair was held in September 1846. William Thomson, at the age of twenty-two, was chosen unanimously by the faculty to be the professor of natural philosophy at the University of Glasgow. David King, William's brother-in-law, was outside the hall where the election occurred and knew the outcome when William's father walked out of the meeting. "A countenance more expressive of delight was never witnessed. The emotion was so marked and strong that I only fear it may have done him injury. . . . " The family celebrated at Knock, and Elizabeth noticed that "during the whole day every now and then a quiet, happy smile" stole over her father's face. Elizabeth ascribed William's unusual composure to the fact that he did "not look in the slightest degree elated."[63] This was strange behavior even for someone as reticent as William Thomson.

Were his doubts rising as to his ability to fill the requirements of the chair? Could he, or was he willing to, bring his teaching down to the "capacity of the ordinary student"? No matter what he thought, the position suited Thomson. He occupied the chair of natural philosophy at Glasgow for fifty-three years until his retirement, and he even remained on after that in spite of blandishments from Cambridge and other universities. Never able to give popular lectures or come down to the level of most of the students, William nonetheless was valued by the university. And when he gained celebrity, the university had more reason to applaud the choice that was made in 1846.

The University of Glasgow was not as prestigious as Oxford or Cambridge, but it had an ancient and proud tradition with a special mission of service to the people of Scotland. It was founded along with St. Andrews and Aberdeen by the Catholic church in the fifteenth century. The three universities were modeled after the University of Bologna, where the diffusion of knowledge and the raising of the general intellectual level were the objectives. Scotland had a special need, the church believed, cut off as it was from the world by the arms of the sea and high mountains. In those remote regions, as Pope Alexander wrote when he founded the University of Aberdeen, lived "a rude, illiterate, and almost barbarous people, who were shut off from the influences of learning and religion."[64]

In 1826, royal commissioners were appointed to inquire into the operation of the universities of Scotland. The commission was one of several that had been convened since the founding of the universities. None had been appointed since 1690, and the occasion of this commission was not to demand drastic reform. Their *Report* recognized and appreciated the distinctiveness of the northern universities. None of the Scottish universities gave entrance examinations or had entrance requirements. While members of the commission believed that some entrance requirements should be introduced, "[they would] consider it to be one of the greatest misfortunes which could be inflicted upon Scotland if any material bar should be opposed to the full participation of the benefits of University education by all. . . . " The need for such a policy was explained in the commission report:

> University education was ardently desired by many whose views were not directed either to the future prosecution of literary studies, or to any of the learned professions, and by many who were intended for the ordinary occupations of the middle ranks of society. Students of this description constituted no inconsiderable proportion of those who attended the Universities. They attended only for the number of years which suited their convenience; they selected the classes which were best adapted to their peculiar views, and often began with those which were the last in the regular curriculum.[65]

When William Thomson testified before another commission in 1877, he opposed any sort of examination for applicants over sixteen. His father was twenty-four years old when he entered the university. A university education should be available to the mature young man who, Thomson said, "has discovered in himself, as is not infrequently the case, a desire and a capacity for higher learning, whether in classics or mathematics. . . . Such men do sometimes in the most wonderful way struggle through the University course, and attain to very high excellence. . . . " The opportunity for an education, William Thomson said, had enabled many to advance their positions in life.[66]

The hard-working students at Glasgow made their contributions to the commercial and industrial life of Scotland and on occasion achieved fame in the arts or sciences. A few professors reached the high ground of success. Adam Smith had come to Glasgow as a student when he was fourteen and finished his education at Oxford.

In 1752, Smith was appointed professor of moral philosophy at Glasgow. He found in the city a worldly and practical atmosphere that he liked. Smith joined the Glasgow Merchants' Club where the members discussed the nature and principles of trade. *The Wealth of Nations,* published in 1776, originated as lectures delivered at the university. Glasgow could not boast about the number of outstanding graduates and faculty as Edinburgh could, but often enough a Joseph Black or a William Thomson studied and taught there indicating that the university was friendly to genius.

It appears from his actions that Thomson did not consider his students different in ability from his father or himself. If they wanted to, he felt, they might learn as much about natural philosophy as he knew. They were, in modern language, educable. There were numerous stories of Thomson's extending his lectures into the realm of advanced research to the consternation of his listeners. One of Thomson's students attested to the fact that the professor's lectures were often too advanced for the level of his class, but Thomson was always interesting. Thomson acted "on the belief that most of his students were on the same plane as himself" and seemed to think that it was impossible for any student in his natural philosophy class to be too dull to understand the lectures. This statement is from a student who was in Thomson's class during 1862–1863.[67]

Thomson's attitude as a teacher reflects an innocence about the capabilities of his students and an extreme unself-consciousness about his own extraordinary capabilities. His feelings for his father colored his estimate of his students. If his father, a self-taught farm lad, could start with nothing and learn all the advanced mathematics as well as be a successful textbook writer, could not any of Thomson's students do the same, if they only tried? Some of his students did go on to do good work, but no one approached Thomson in their accomplishments. Thomson was not going to be a wrangler maker or the Glasgow equivalent. Nor did he have Hopkins's talent for directing a student's studies in a way that would conduct him to the heights from which he could glimpse the territory of original research. Such talent enables the teacher to let himself down to the student's level and gradually urge him to an advanced level of thinking. Teaching at the introductory level of science in a university where science is part of a general studies program requires that the teacher use an intellectual anchor to slow his thinking down to the student's rate of comprehension. A child prodigy himself, Thomson did not understand this simple truth and was unable to make the necessary adjustment. Even when he introduced his teaching laboratory, the first of its kind in Great Britain, students gained experience by assisting in the conducting of original research. This was far better, perhaps, than cookbook laboratory experimenting, but a type of education that suited only a select few.

But the teaching did offer advantages for Thomson whose whole energy was always directed toward original research. There was the necessity of explaining the subject without using advanced mathematics. William's constant challenge was to lecture in physical-geometrical terms. The preparation for his first lecture brought physical and experimental matters to mind and started him off on a new vein of research. The evening before the lecture, Thomson was still struggling with what to say in his introductory remarks. And the pressure, as great in its way as preparing for the Tripos, forced his mind into another channel:

> Glasgow, October 31, 1846   11:45 P.M.
> I have this evening (in the middle of my work finishing an introductory lecture) after thinking on Faraday's discovery of the effects of magnetism on transparent bodies and polarised light, been recurring to my idea (which occurred to me in the May term) which I had to give up, about magnetism and electricity being capable of representation by the straining of an elastic solid constituted in a peculiar way. I *think* the following must be true.

There followed two pages in which he described his new ideas. He finished with the regret that "I am not at all sure of any thing I have written just now, but I want to put it out of my head, as I shall have no time to spare, during the session."[68]

Turning the page of the diary, one is surprised to see that he did find time to write on 21 November at 11:30 P.M., and on 28 November at 10:15 P.M. he wrote: "I have at last succeeded in working out the *mechanico-cinematical* (!) representation of electric, magnetic, and galvanic forces." The awkward term "mechanico-cinematical" was coined to describe William's mechanistic and dynamical analogy for electrical force. On the twenty-ninth, at 2:45 A.M., a paper for the *Cambridge and Dublin Mathematical Journal* was finished.[69] The paper, titled "On a Mechanical Representation of Electric, Magnetic, and Galvanic Forces" represented Faraday's experiments with polarized light by a new mathematical analogy and used G.G. Stokes's mathematical treatment of equilibrium in an elastic solid as the basis for the analogy.

As Thomson realized later in life, the professorship at Glasgow was ideally suited for continuing his work in science. In a remarkable way, it combined his physical and mathematical bent. There were a number of reasons for this suitability. First, there was the physical laboratory begun at once for his own research and expanded later to enable students to participate in original research. Then, there was James White, the optician and instrument maker who first provided the tools for research and later became the factory for Thomson's patents. Even the lectures served their purpose for giving simple physical explanations, and on several occasions, they became a stimulus for new lines of inquiry.

What that stellar nineteenth-century scientist contributed to the academic climate at Glasgow was something unique and of use to only a handful of students. The truth was that Thomson's lectures were not suited to the ordinary student at Glasgow. His legendary lecturing provided material for many an anecdote. In "Lord Kelvin, A Biographical Sketch," J.D. Cormack reported the following about William's lecturing in his later years:

> In the two lectures per week, which he delivers to the junior class, he contrives to touch on most of the important points of his subject, passing from one to the other with a rapidity bewildering to lazy or ill-trained students who prefer the beaten track. On such members his digressions, often the most fascinating as well as the most valuable part of his lectures, are, alas, almost lost. His master mind, soaring high, sees one vast connected whole, and, alive with enthusiasm, with smiling face and sparkling eye, he shows the panorama to his pupils, pointing out the similarities and differences of its parts, the boundaries of our knowledge and the regions of doubt and speculation. To follow him in his flights is real mental exhilaration, a revelation and a keen delight to the better students, and the eyes of even the dullest lift at times from a seeming weary endless road and see the glory of the view.[70]

Even the duller students looked forward to Thomson's classroom demonstrations. In one of these, he determined the velocity of a bullet by firing an old Jacob rifle into the heavy wooden bob of a pendulum. There was great excitement, and some amusement, to watch the professor take careful aim and fire. Before the smoke had cleared, Thomson was at the board doing the calculations.[71]

CHAPTER SIX

From Statical to Dynamical

On 1 November 1846, William Thomson began his long career as professor of natural philosophy at the University of Glasgow. By 5 November, he had given two lectures and an oral examination. His introductory lecture, he wrote to Stokes, "was rather a failure as I had it all written out and I read it very fast." Stokes tried to give the harried new professor some perspective: "When once your first course is completed you will not find it such hard work." His reassurances were borne out, and Thomson's inaugural year ended in triumph despite the tragedy that struck in the middle of the session.[1]

William's younger brother John died suddenly at the age of twenty-one. He had succumbed to a disease he contracted at the hospital where he was training for his medical degree. J.D. Forbes wrote that "it is a fresh instance of the melancholy loss amongst our best rising young medical men by that horrid disease." Consequently, John's death cast a pall over the rest of the session and William was anxious to get away to Cambridge. In March he told Stokes to expect him in May. Immediately after the end of the session both he and Dr. Thomson left for Cambridge, where William had gotten his father a room at Peterhouse for his two-week stay. Father and son also made a trip to London.[2]

When Dr. Thomson returned to Glasgow, their correspondence resumed, and on 22 June 1847 William informed him he was "enjoying [himself] very much here at present, as it is remarkably quiet, and I really have for a week found a little time for myself. I have been getting out various interesting pieces of work, along with Stokes, connected with some problems in electricity, fluid motion, &c. that I have been thinking on for years, and I am now seeing my way better than I could ever have done by myself, or with any other person than Stokes."[3] Thomson's unsolved problems always remained latent in his memory, and he retrieved the unfinished work quite easily from his mathematical notebook.

During a week in London Thomson visited Cayley and Faraday.[4] Usually his trips to London included a visit to the Royal Institution to hear Faraday lecture and to talk with him afterward. Undoubtedly, their conversations were taken up with the details of Faraday's experiments; they had revealed some new characteristics of electrical and magnetic phenomena, thereby showing previous theory as inadequate. Thomson had an opportunity to ask Faraday about his experiments with the effect of magnetism on polarized light. These experiments established a new relationship between electrical and magnetic force. Faraday had passed polarized light through a transparent object and found that when the object was placed in a magnetic field, the plane of the polarized light was rotated. The stronger the magnetism, the farther the plane was rotated. This discovery added to the long list of effects of magnetism and electricity, which included attraction and repulsion between magnetic substances, as well as between magnetic and electromagnetic forces, the induction of magnetism from one body to another, the induction of an electric charge from one body to another, the induction of an electric current in a wire by means of electromagnetism, as well as electrochemical effects, the production of heat by electricity and the production of electricity by heat. Now Faraday had added light to the complex of electromagnetic phenomena.

For scientists such as Snow Harris, the multiplicity of properties exhibited by electricity and magnetism caused them to despair of finding a theory that incorporated everything that was known about the mysterious fluids. A theory had to explain the different actions of electricity and magnetism in order to be accepted as anything more than provisional. Green offered a theory that included only the exertion of the force of attraction or repulsion, making an analogy with gravitational force. Faraday's magnetism and polarized light experiments seemed to indicate a force being exerted through the transparent body on the light. Thus a theory dealing with electric and magnetic force had to include an explanation of his results. Some theorists were chagrined at the burden Faraday's results had placed on them, but for Thomson each of Faraday's discoveries gave another glimpse of the character of the imperceptible forces. Faraday's experiments were an aid, not a bane, to Thomson's efforts to develop an inclusive theory. But he pictured these results in a better way when they were expressed mathematically: Faraday's physical explanation, using the idea of lines of force, was no more a help to Thomson than it was to the French mathematicians.

The most recent phase of Thomson's development of electrical and magnetic theory began on the eve of his first lecture to the natural philosophy class in the fall of 1846 and concluded with his discussions with Stokes in Cambridge. Thomson's coined word "mechanico-cinematical" for representing static electric, magnetic, and electro-magnetic forces conformed to his idea of an analogy with an elastic solid. The twisting force of Faraday's magnet which turned the plane of polarized light suggested to Thomson the force exerted to twist an elastic body, for example a cube of rubber. He represented the force exerted by a bar magnet by the amount of force required to twist an elastic solid out of shape.[5] Therefore, by means of analogy Thomson substituted problems in elastic solids for those dealing with electric and magnetic force. When Faraday talked about the induction of an electrical charge from one body to another, he said that it occurred through the action of contiguous particles, and his experiments indicated that it lay along curved lines. This physical explanation, which spoke of the action between particles, was unacceptable to Thomson because there was no way of reducing it to mathematical terms. Thomson's analogous idea of an elastic solid had the advantage of following Faraday's belief in direct action of forces through contiguous particles (as opposed to action at a distance).

The physical analogy between electric and magnetic forces, and the strain in an elastic solid had been the basis for Thomson's article "On a Mechanical Representation of Electric, Magnetic, and Galvanic Forces," written in November 1846 and published in the *Cambridge and Dublin Mathematical Journal*.[6] The mathematics for equilibrium in elastic solids under stress had been given by Stokes, and Thomson demonstrated that particular solutions of Stokes's equations represented the three forces: static electric, magnetic, and electromagnetic. The relations between the elastic and electric equations were geometrical in that the locus of force was given by differential equations, which indicated the direction and magnitude of the different forces.

This was a mathematical metaphor. The symbols and algebraic relationships in the equations not only represented the forces produced by strain in an elastic solid, but the equations themselves also provided a means of solving for the electric and magnetic forces. The physical analogy was a limited concept, but the mathematical metaphor suggested other relationships. That was the reason why Faraday's discovery of the effect that magnetism had on polarized light did not unsettle Thomson's preconceptions as it did those of other scientists. For Thomson, Faraday had supplied another piece of the puzzle linking together all those forces—analogously, for the time being. The mathematical metaphor was the way theoreticians were to understand the relation between all forces or energy. It was through the metaphor that mathematicians were able to accept the direct relationship, that is, the interchangeability of all energy.

For Thomson, now, nothing seemed impossible in the relation between forces. He proposed that an experiment be tried to induce magnetism in a steel needle by passing polarized light around the needle. Admitting that the experiment sounded foolish, he said: "After Faraday's various discoveries, hardly anything should appear too improbable *a priori* to be not worth trying. . . . "[7]

An exchange of letters between Thomson and Stokes that began in the spring of 1847 concerned another of Thomson's physical analogies. The attraction of a bar magnet, he thought, might be represented by water flowing in a closed tube. If the ends of the bar magnet were imagined to be bent together, then the velocity and direction of the motion of the water were analogous to the intensity and direction of the magnetic force. Had he written the mathematics correctly?, he asked Stokes. Thomson's mathematical metaphor suggested several possibilities to Stokes; one of them he saw as the same as the problem of determining the permanent temperature of a homogeneous solid. The mathematical equations which they were both using, Stokes wrote, might be useful "in the universal calculation of problems in heat and electricity. . . . " Stokes wanted to use the equations for problems in fluid motion.[8]

The parallel paths that Thomson and Stokes were following in their theoretical work were, in 1847, taking a turn toward convergence. Thomson's idea for representing magnetic attraction as well as his realization that Faraday's work opened the way to combining static and current electrical force had its counterpart in Stokes's work on hydrodynamics. In his "Report on Recent Researches in Hydrodynamics" to the British Association in 1846, Stokes described the relation between the study of hydrostatics and hydrodynamics so well as to serve as a standard reference work. In the report, he argued that the fundamental hypothesis in hydrostatics was the same as that used in hydrodynamics.[9]

With Stokes's mathematical treatment of elasticity and hydrodynamics supplying so much necessary information to Thomson at the precise moment when it was filling in the connecting links in Thomson's thinking, it was an undue modesty in Stokes to apologize for taking up Thomson's time, perhaps uselessly. "I do not know," Stokes wrote, "but that all the ideas in this letter may be quite old to you. If my letter does you no good it will at least do you no harm further than taking a few minutes to read." This was said more than once during their interchange. But Stokes did know that their combined thinking was supplying some very new ideas. "What an intimate relation there is between the mathematical considerations which are applicable to heat, fluid motion, and attraction!"[10]

Thomson's collaboration with Stokes was his first, longest lasting, and most significant relationship in the way it expanded his range of thought. In theory, too, as in Faraday's experimenting, almost anything seemed possible, but the cautious, disciplined mathematical thinking of Stokes kept Thomson's speculations within bounds. Stokes's mathematical approach to hydrodynamics, through which he envisioned the rational transition from hydrostatics to hydrodynamics, was a way of conceptualizing dynamical force in the familiar terms of statics. The study of hydrodynamics became for Thomson a fundamental research tool that was seemingly applicable to all natural phenomena. As he wrote to Stokes years later, the study of hydrodynamics vied with electricity as the key to nature.[11]

In 1847, Thomson wrote the first of his articles on the mathematical theory of hydrodynamics. Over the next two years, he and Stokes published a series of articles in the *Cambridge and Dublin Mathematical Journal* that explored the basic concepts of hydrodynamics.[12] Unlike electricity, hydrodynamics was visible in many everyday occurrences from ocean waves to the goings on in one's cup. The observations served as a source of amusement for Thomson. From Paris, he wrote to Stokes that he had seen a fine instance of elasticity in an incompressible liquid. . .

> on a cup of thick "chocolat au lait." When I made the liquid revolve in the cup, by stirring it, and then took out the spoon, the twisting motion (in eddies, and in the general variation of angular velocity on account of the action of the spoon overcoming the inertia of the liquid, and the friction at the sides) in becoming effaced, always gave rise to several oscillations so that before the liquid began to move as a rigid body, it performed oscillations like an elastic (incompressible) solid.[13]

Through his lifelong correspondence with Stokes, Thomson received an immeasurably important stimulus to his thought. Stokes was a fount of new ideas and a testing ground for Thomson's speculations. Like Thomson, he was interested in every new idea in science. He was a live stimulus, unlike Green and Fourier who, now dead, could be only unresponsive reservoirs of ideas. In 1847, the Thomson-Stokes exchange explored the interconnections of force or energy:

From Statical to Dynamical

College, Glasgow
April 7, 1847

My dear Stokes,

Many thanks for your letters, which have given me plenty of matter for contemplation, in subjects with which I have long been interested. There is a great deal which I would like to say about them, but I do not know where to begin. . . .

I have been for a long time thinking on subjects such as those you write about, and helping myself to understand them by illustrations from the theories of heat, electricity, magnetism and especially galvanism; sometimes also water. I can strongly recommend heat for clearing the head on *all* such considerations, but I suppose you prefer cold water.

You will be quite delighted if you read a few pages or even a few words of a paper of Gauss's . . . (Taylor's Scientific Memoirs, Vol. II. p. 230), by which you will be supplied by a flood of illustrations for water. For instance, for motion in 2 dimensions, in an infinite space, round circular cores. . . . [The mathematical treatment of his illustration, Thomson claimed, was useful in dealing with heat, hydrodynamics, or magnetism depending on the interpretation of the lines in his diagram. But he was not sure that it was possible to find a solution for the general equations.] I think I see my way now however, and if I am right, I shall have an immense number of new problems in heat, electricity, and magnetism, to think on, and I shall try to make a proof suitable for cases of attraction. . . .

Yours very truly,

William Thomson[14]

Thomson's new broadened outlook, a consequence of Stokes's hydrodynamics, offered him a way of incorporating some ideas which up to then he had resisted. One of these new ideas was Ampère's theory of circulating currents. In a paper titled "On the Electric Currents by which the Phenomena of Terrestrial Magnetism may be Produced,"[15] given before the mathematical and physical section of the British Association at Oxford in 1847, he revealed how far and in what ways he had departed from his statical views.

Terrestrial magnetism was an idea which Thomson used for developing his mathematical theory of electromagnetism. The mathematical theory, he said, was a combination of Green's theory, Ampère's theory, and a third mathematical theorem in which fluid motion was a physical instance. Green's theorem stated that the action of any body, whether it was exerting a gravitational, magnetic, or electrical force, might be represented by means of the distribution of matter over the surface which will produce exactly the same force as the body itself produced. The theory conceived of three types of imaginary matter, gravitational, magnetic, and electric, and the way this matter was distributed over the surface of the body determined the magnitude and direction of the force. Thus by means of integral calculus, the distribution of the imaginary matter was summed up and the total was equal to the force. Thomson said that the force of terrestrial magnetism might be found using Green's theorem, by finding the particular distribution of imaginary magnetic matter that would produce the observed effects of the earth's magnetism. However, he cautioned that although Green's theory was of great theoretical value—that is, it had mathematical use—it was not the expression of physical fact.[16]

Thomson also called Ampère's theory "merely theoretical," although Ampère proposed it as a physical fact.[17] This theory viewed all magnetism as being caused by circulating electric currents. In particular, Ampère thought of the earth as a large electromagnet which caused terrestrial magnetism. Although Thomson thought Ampère's idea was closer to the physical reality, he still called it "merely theoretical" since he was unable to conceive of electric currents going around molecules of matter.

The question Thomson wanted to answer was stated in the title of the paper, "On the Electric Currents by which the Phenomena of Terrestrial Magnetism may be produced"—not "produced," but "may be produced." If it were possible to ring an object the size of the earth with wires carrying electric currents, could there possibly be some pattern of wires and magnitude of current flowing in them that would produce the same effects as the actual terrestrial magnetism? By using the three mathematical theories of Green, Ampère, and Stokes, Thomson was able to find the equations that described the distribution of the hypothetical closed electric circuits.

The statements in this paper make firm distinctions between what was merely theoretical and what was a physical fact. By the term "merely," Thomson was not denoting a lesser value or importance to ideas such as Green's and Ampère's, but he wanted to indicate that the theories were to be taken only as a hypothesis about physical reality, which was useful in deriving mathematical theory, and that one should not be misled into thinking that Green's imaginary matter had any kind of existence outside of the imagination.

The objective of scientific research, Thomson felt, was to be able to give a concise description of physical reality. But so little was known (Thomson would add "in the present state of our knowledge") about terrestrial magnetism at that time that a merely theoretical approach was necessary. This approach resulted in the mathematical description, which in turn guided experimental research. By these steps, Thomson was confident, physical reality would be known directly.

Twenty-five years later, Thomson would admit that at the beginning of the summer of 1847 events had gradually brought him to the point where he was prepared to accept Ampère's theory as physical reality. The whole course of events, he said, had caused him to abandon one by one all statical conceptions of force for dynamical ones. Stokes's theoretical transfer of hydrostatics to hydrodynamics was the first step. Another step was made that summer when Thomson first heard Joule's idea about the dynamic nature of heat.[18] Green's theory was an analogy between gravity and electrical attraction, and since gravity was caused by a dormant mass his analogy was statical.

By use of analogy Thomson was able to inform new areas of study such as electromagnetism by ideas from the better understood areas of elasticity and hydrodynamics.[19] Thomson was not able to use analogy with a theory such as Ampère's whose physical reality he could not accept. Therefore, Thomson assumed Ampère's equations to be a mathematical metaphor. Ampère's equations made reference to circulating currents, which were no more real to Thomson than saying that the prow of a ship is a plowshare when it "plows" the seas. But taking Ampère's theory to be a mathematical metaphor for terrestrial magnetism was stimulus to Thomson's imagination. Because of the level of abstraction, the mathematics of Thomson's solutions was able to combine Green's theorem (an analogy with gravity), Ampère's theory (an identity with electricity), and Stokes's theory (fluid motion) to represent terrestrial magnetic force. Only in mathematics was it possible to combine such *physically* disparate phenomena to create a metaphor.

Once Thomson accepted Ampère's view as metaphor, his imagination was freed to consider many new possibilities related to this dynamical view. The view was dynamical because it supposed that magnetic force was due to the motion of an electric current so that change in the magnetic force could be designated by a change in the magnitude of the current. According to the Green analogy, change occurred from one state of static equilibrium to another by adding or subtracting imaginary matter. The image was less suggestive for change in force or energy, and, after all, the whole concept of energy involves the idea of change. Seeing the mathematical metaphor between all forms of energy, Thomson moved freely from the study of one form to another, and gradually, as he recognized twenty-five years later, the idea became impressed on his consciousness that all phenomena were in reality dynamic, and mathematically he had a dynamic metaphor.

In the terrestrial magnetism paper, Thomson first made use of a model based on the dynamic metaphor. He had, he wrote, constructed an electromagnetic model of the earth after Ampère's theory.[20] According to this model, terrestrial magnetism exerted a force as if the earth were covered with wires with current flowing through them. Where Green's theory was a static analogy, the model was a dynamical representation—all "merely theoretical," but highly suggestive to the imagination. Thomson's progress from the use of analogy to the dynamic model was via the mathematical metaphor. Although not representing the thing itself, the metaphor actually led him to think in dynamical terms and over the years it became a habit of mind.

Along with his discussions with Stokes, work was going so well in Cambridge that Thomson wanted to stay quietly settled for a time. His plans were to go to Switzerland later that summer after a brief stay in Paris. He took a few days off from his research for the trip to London, but he was now tempted to curtail all other travel for the quiet of Cambridge. The thought was a passing one; on 22 June 1847 he wrote in his diary that he'd been thinking about "saying something about representing terrestrial magnetism by superficial currents . . . as well as about images" at the meeting of the British Association in Oxford. The next morning, he left for Oxford, where he gave the two papers he was thinking about.[21]

Ludwig Fischer twitted Thomson with news that Sir David Brewster's daughter would not be in Oxford: "I am afraid you will be greatly disappointed. Your charmer, the accomplished poet . . . will probably not be there. . . . " A few days later, Fischer wrote, "You must be delighted to hear, that after all Miss Brewster and the Misses Lyon are going to Oxford, and as you are determined to be charmed with the accomplished daughter of the great St. Andrews philosopher. . . . " William informed his father that he enjoyed the week in Oxford very much, but there is no evidence available that a young lady contributed to William Thomson's enjoyable week.[22]

The meetings of the British Association were open to all cultivators of science, bringing together the amateur and the serious researcher, as well as the experimentalists and the mathematical theoreticians. Sections were designed for those with special interests, but by moving about freely between section meetings one easily found what was of primary interest at the moment, and there was an opportunity to meet people who shared one's interests. Thomson met Professor Nichol, James D. Forbes of Edinburgh, Faraday, Sir William Hamilton, and he had a long amiable talk with W. Snow Harris.[23]

After hearing Thomson's paper titled "On Electrical Images," Forbes asked Thomson to give it as his "Maiden Essay" in the *Transactions of the Royal Society of Edinburgh*. (Thomson was elected a fellow of the society on 1 February 1847.) Forbes repeated his invitation a few weeks later. On 15 August, Thomson wrote his father that "[he was] trying to get on with a memoir on Electrical Images." Sir David Brewster also asked him to give the essay on images to the Royal Society of Edinburgh.[24] On 19 August, Thomson wrote in his diary:

> I have this morning, at last, determined what to do in the way of publishing results. In my essay for Edinburgh Royal Society I shall give the principle of images as it first occurred to me, in connection with general theorems. In the January No. of the Cambridge and Dublin Mathematical Journal I shall give the geometrical method (see letter to Liouville for 8 October 1845) indicating the other, as being more suitable after something shall have been published in the Journal on the elementary theory. I may perhaps free the paper in the Journal entirely from symbols of differential and integral calculus and do the theory of distribution on spheres "by Newton."[25]

As it turned out, the "Maiden Essay" was not "On Electrical Images." It was "An Account of Carnot's Theory of the Motive Power of Heat, with Numerical Results deduced from Regnault's Experiments on Steam."[26]

One of the papers Thomson heard at the Oxford meeting was "On the Mechanical Equivalent of Heat, as determined by the Heat evolved by the Friction of Fluids," given by James Prescott Joule. Joule's premise, stated in 1843, was that heat was not a substance, that is, caloric, as commonly assumed, but a "*state of vibration*";[27] that heat was convertible into mechanical force; and that this conversion occurred at a fixed ratio. The fixed ratio between heat and mechanical force he called the mechanical equivalent of heat.

Like Faraday, Joule was an experimentalist who developed a theory to explain the results of his experiments. He was twenty-eight years old at the time of the Oxford meeting and had been experimenting for over eight years. His laboratory had been established in order to test the feasibility of using the newly invented electric motor in the brewery by comparing its efficiency with the steam engines used in the plant. By converting the "duty" of a steam engine and that of an electric motor to the same standard of measurement, the foot-pound, Joule found that burning a pound of coal to supply heat to the steam engine was much more economical than an expensive pound of zinc consumed by the electric battery which supplied the electric motor. Wondering why the battery was so inefficient, Joule measured the heat lost by battery electrolysis and the heat lost in the conductors of the electric motor when current passed through them. The heat losses robbed the electric motor of power; without that handicap it would have been a better competitor for the steam engine.

To find a common standard of measurement, Joule converted all forms of energy—heat, electrical, and chemical—to the common denominator of mechanical power, the foot-pound. Abandoning the limited question of whether the electric motor or the steam engine was more economical, he turned to the more general idea of convertibility of energy. He considered electricity, which produced heat in a conductor and a mechanical force in a motor, as "a grand agent for converting heat and the other forms of mechanical power into one another."[28]

Joule turned from engineering to experimental science; in letters to the *Philosophical Magazine,* in addresses to the Manchester Literary and Philosophical Society, and in papers before the British Association he described the variety of experiments that repeatedly demonstrated that there was a fixed ratio, or equivalent, between mechanical force and heat. Of all the energy conversions, these were the two specific ones he was most interested in. For example, in 1845 he found the equivalent by measuring the force applied to a paddle wheel immersed in water and the temperature rise of the water—a measure, he said, of the heat created by the paddle wheel's motion. In 1847, he wanted to tell the British Association about the improved accuracy of his experiments in determining the mechanical equivalent, but the secretary only allowed him to give a summary without giving the observational data.

Joule had not been very successful in winning adherents to his theory concerning the conversion of heat to mechanical motion because it appeared to conflict with the accepted idea that heat was a substance. At Oxford he found at least two interested listeners, Stokes and Thomson, who wanted to know more about his ideas. Joule was delighted and gave Thomson two of his previously published, but unnoticed, papers.[29]

Why was Thomson interested in Joule's work? He was not certain whether Joule had the foundation of an important general theory or a specific experimental exception not yet covered by the accepted caloric theory. Fresh in Thomson's mind were the analogous relations between different forms of energy. The topics of the two papers he gave at the association meeting concerned analogies. The paper on electrical images dealt with the analogy between electricity and light, while in the one on terrestrial magnetism Thomson chose to think of an analogy between the earth's magnetism and electromagnetism. Thomson had employed a number of analogies for electricity—elastic solid, fluid flow, heat, gravity, and light—so that Joule's data on his purported conversions suggested an experimental counterpart to Thomson's theoretical analogies. Were Thomson's analogies realistic? Was the convertibility of one form of energy to another the materialization of his imagery? Thomson's equations, his mathematical metaphors for the conversion, suggested a relationship with Joule's experimental conversions. But Thomson, like the other scientists there, found the obstacle to accepting Joule's theory of mechanical equivalence to be the well-established caloric theory which was the foundation of so much in thermodynamics.

Thomson sent copies of Joule's papers to his brother James, who would see the engineering implications of the new theory. However, William warned James: "Joule is I am sure wrong in many of his ideas, but he seems to have discovered some facts of extreme importance, as for instance that heat is developed by the friction of fluids in motion. . . . " Thomson thought that Joule's papers "[would] astonish [James]," and he thought there were some serious flaws in Joule's results. James had doubts also, but had to admit that Joule's ideas were unsettling to anyone who accepted Carnot's theory about the materiality of heat.[30]

The ambivalence that Thomson felt about Joule's claims was exhibited in his paper titled "On an Absolute Thermometric Scale Founded on Carnot's Theory of the Motive Power of Heat, and Calculated from Regnault's Observations," published a year later in June 1848.[31] The limiting factor to the accuracy of thermometers was the lack of uniform expansion over a wide range of temperatures of the fluids (water, mercury, or alcohol) used in the thermometers. An absolute scale based on a mathematically precise relationship as described by Carnot gave the basis for an exact scale without any limit to the accuracy which might be achieved. The reliance on the varying expansive qualities of fluids was a relative temperature scale.

The principle on which Carnot's theory was based came from a theoretical examination of the steam engine. The mechanical effect or work produced by a steam engine depended on three factors: the quantity of heat supplied by the steam boiler, the temperature at the beginning, and the temperature at the end of the steam engine's cycle. It followed from the theory that the perfect engine, as defined, was one where the application of mechanical force to the engine operating in reverse raised the quantity of heat at the outlet to the temperature of the heat, caloric, at the beginning of the cycle. In other words, the steam engine produced work by means of a quantity of caloric falling from a higher temperature to a lower temperature. The theory not only isolated the factors necessary for the operation of a steam engine but, by defining the perfect engine, it also gave a standard of measurement for determing the efficiency of any engine.

Carnot's theory was the established theory in 1848. In the twenty-four years since its publication, it had been accepted by engineers as the standard way of describing the operation of the steam engine. Regnault's research for determining the constants to be used in applying Carnot's theory was being supported by the French government in its efforts to aid French industry. How completely Carnot's theory had been absorbed was indicated by the scarceness of his book. Thomson was unable to find a copy of Carnot's *Réflexions sur la Puissance Motrice du Feu* when he was in Paris in 1845. He did not obtain a copy until 1848.[32] He learned about Carnot through reading Clapeyron's book, which was an experimental interpretation of Carnot. The original statement of a scientific theory becomes less and less available as it is interpreted and reinterpreted by other scientists. No one needs to read the original formulation of the theory unless he wants to challenge the basic premise of the theory.

Thomson's 1848 paper was important because it examined the suppositions involved in Carnot's theory. The absolute temperature scale was based on Carnot's assumption that the mechanical effect of a steam engine was produced by the *transmission* of a quantity of caloric from a higher temperature to a lower one. The definition of an absolute degree (later named "degrees Kelvin" in honor of Thomson) was the amount that a unit of caloric had to fall in order to produce a unit of work. It was assumed that nothing happened to the caloric in passing through the steam engine; it was taken for granted, as an axiom, that heat was not convertible to anything else.

The analogy most often used was the waterwheel, where the power was derived from the water falling against the wheel in its descent from a higher point to a lower one downstream. The output of the waterwheel was determined by the quantity of liquid and the space through which it fell; in the steam engine only the quantity of

caloric and the fall in temperature matter. There was no more reason to believe that caloric was *converted* to something else during its
passage through the steam engine than to believe that water was
transmuted to some imponderable substance in passing through the
waterwheel.

Thomson wrote in his paper, "the conversion of heat (or *caloric*)
into mechanical effect is probably impossible, certainly undis-
covered." A footnote claimed that this opinion was almost universally held, but that the contrary opinion was insisted upon by Mr.
Joule of Manchester who had made some "very remarkable discoveries" with reference to "*generation*" of heat by motion. However, Joule
had not been able to do the reverse—convert heat into motion and
until he showed by experiment that heat was convertible into motion
his theory was unproven. There was much about these questions,
Thomson confessed, that was still a mystery.[33] Thomson resisted the
implications of Joule's experiments, but his footnote betrayed his
first doubt about the existence of caloric.

Joule first saw Thomson's article "On an Absolute Thermometric
Scale Founded on Carnot's Theory of the Motive Power of Heat,
and Calculated from Regnault's Observations" in the October 1848
issue of the *Philosophical Magazine*. (Why didn't Thomson send Joule
a copy of his paper? Was it due to his unwillingness at all times to
offend anyone?) Joule wrote Thomson that he knew his idea was a
valuable contribution to science even though it was based on a wrong
premise, Carnot's theory, and that Thomson's absolute temperature
scale would be useful even when Carnot was finally found to be
wrong. Joule added: "There is one point mentioned in your paper
especially interesting to me namely that the possibility of *converting
heat into mechanical effect* is not admitted." He granted that there were
one or two points still unsettled but thought he had proven the
convertibility of heat into mechanical force in his experiments. As a
further confirmation, he was planning a series of experiments.[34]

Thomson replied with a nineteen-page letter. He confessed an
inability to answer Joule's exceptions at the moment, for he wanted a
good deal longer to collect his thoughts on the subject. He hoped for
an "ultimate reconciliation of our views." Thomson wrote that he
planned to conduct his own experiments and was also waiting for
the results of Regnault's latest experiments. "I am quite aware,"
Thomson wrote, "of the importance of the objection you adduce to
Carnot's Theory . . . which I admit in its full force. . . . I have never
seen any way of explaining it although I have tried to do so since I
first read Clapeyron's paper; but I do not see any modification of

the general hypothesis, which Carnot adopted in common with so many others, which will clear up the difficulty."[35]

When Thomson finally obtained a copy of Carnot's book, he wrote an article titled "An Account of Carnot's Theory of the Motive Power of Heat; with Numerical Results Deduced from Regnault's Experiments on Steam," which he read to the Edinburgh Royal Society in January 1849.[36] His purpose was to examine Carnot's basic assumptions stripped of any later interpretations in order to determine which were the essential features of Carnot's theory that made it so useful, to show how Regnault's experimental data was a verification of the theory, and, finally, to apply the theory by means of Regnault's data to calculate the efficiency of steam engines already installed and operating in different parts of the world. The underlying question in Thomson's paper was: Was Carnot's theory sound? To find that out he examined its reasonableness (in particular, whether the assumptions were reasonable) and then tested the theory in the region it was designed for—that is, to measure the efficiency of steam engines.

What are the main properties of heat, Thomson asked, that pertain to the operation of steam engines? The quantity of heat can be measured by the temperature of a body that contains heat. Heat raises the temperature of bodies. It can also increase the volume of a body through expansion. Where there is nothing to resist an increase of volume, as in the open air, heat simply increases the volume. If the expansion is resisted, when the carrier of heat is confined, as it is when steam is injected into the cylinder of a steam engine, then the resistance to expansion produces a mechanical effect. Temperature, volume, and mechanical effect are the three measurable factors that account for the operation of a steam engine.

In operation, some of the heat in a steam engine raises the temperature of the steam, some increases the volume, and some of the increased volume is resisted; the result is mechanical effect. In a perfect engine, Carnot reasoned, reversing the process by applying mechanical effect would overcome the resistance to the change in volume and restore the steam to the same volume and temperature that it had at the start of the cycle. A perfect engine was one which gave maximum mechanical effect from a given quantity of heat. What is the most work that can be obtained from a steam engine? It is the maximum mechanical effect that can be used, in a reversed engine, to restore the heat to its original temperature with the volume of the piston the same as at the start. The fundamental assumption was that heat, being a substance, does not change in quantity when passing through the engine; it only falls in temperature and, in doing so, expands the steam within the cylinder and produces the mechanical effect. If the heat changed in quantity, there was no basis, according to Carnot, for measuring the mechanical effect of the reversed engine because there was no way of telling how large a quantity of heat disappeared.

The analogy with the waterwheel, which was often used in describing the Carnot cycle, shows how important the assumption was that heat was not converted to anything else. In the waterwheel, the water (by analogy, heat) fell from a higher point (by analogy, higher temperature) to a lower point and produced mechanical effect in the fall of water by turning the wheel. To measure the maximum mechanical effect of the waterwheel, the total amount of water (heat) at the downstream end might be collected in a container and carried back to the original height. The amount of effort required to carry the water back to its origin was a measure of the maximum mechanical effect obtainable from the wheel. All that had to be ascertained was that *all* of the water was collected at the tailrace of the waterwheel. If some of the water had been dissipated or transformed, then the measurement of the effort required to restore it to its original position would be inaccurate by the amount of the transformation of water. But of course water did not transform into anything else, and by analogy neither did heat—everyone believed that except Joule.

The experiments that Joule had conducted with electricity and also with friction of fluids seemed, Thomson admitted, to refute the axiom on which Carnot had built his theory—that heat cannot be generated. Joule claimed to have shown that the heat caused by an electric currect was not transferred from the electric generator, but was generated in the conductors themselves. Joule raised a serious objection to the assumption of the materiality of heat, Thomson admitted, but then he went on to challenge Joule's experimental technique. Perhaps the heat in the conductors had been transferred from the magnets of the generators, and Joule had not demonstrated that the magnets had not lost heat.[37]

Although Thomson expected, in 1849, that Joule's experimental results would be reconciled with Carnot's theory, he chose Carnot as the only acceptable explanation. To his own question, "What is the precise nature of the thermal agency by means of which *mechanical effect* is to be produced, without effects of any other kind?", he answered, *"The thermal agency by which mechanical effect may be obtained, is the transference of heat from one body to another at a lower temperature."*[38]

Joule's claim that heat was a form of motion and not a material substance cast doubts on the very foundation of the theory of heat. The perplexity left no alternative, Thomson thought, than to design experiments "either for a verification of Carnot's axiom, and an explanation of the difficulty we have been considering; or for an entirely new basis on the Theory of Heat."[39] There were no new experiments. Yet on the basis of what he already knew, Thomson changed his mind just one year later, in 1850. In an ultimate acceptance of Joule's theory, Thomson collaborated with him in a series of experiments conducted between 1852 and 1856.

The traditional statical view of the materiality of force represented the different forces of nature by different kinds of imponderable matter. The new dynamical view emphasized force as the result of the motion of ordinary matter. The traditional view explained the differences between heat, electricity, and magnetism as being consequences of the peculiarity of the type of the substance—the imponderable; the relationship between forces, or imponderables, was understood by Thomson by means of analogy. In the dynamical view, all force or energy was understood to be different forms of an underlying field of energy, and all forces were convertible one into the other. The search was for a device that would convert force from one type to another as in the steam engine or the electromagnet. The relationship between forces, in the dynamical view, was denoted by numerical equivalences of the transformation of different forces. Joule called his equivalent "the mechanical equivalent of heat," that is, the equivalent amount of mechanical force that was obtainable from a unit of heat. The concept of analogy was still relevant. However, in the dynamical view, it was based only on numerical equivalences of kinds of energy, not on fluid or mechanical models. And research in dynamical science was to consist of a search for constants of equivalence that would relate one force to another. The study of the different types of force was generalized by use of these constants of conversion that were used in equations for describing all conditions for conversion. Later, the Thomson-Joule experiments would contribute to the establishment of these constants.

Events in Thomson's life that occurred between 1847 and 1849 marked his growing maturity; besides undergoing a fundamental change in his thinking, he cast off some aspects of his youth and fixed some of his intellectual qualities so that they were more determining for the direction of his future. All aspects of his life contributed to his thinking, or he managed to surround any disturbing intrusions with a concretion the way an oyster surrounds a foreign object within its shell.

After the meeting of the British Association, Thomson set out on a trip to Switzerland for a relaxing vacation. First, he stopped in Paris and placed an order for 1,000 francs of acoustical equipment for his lectures at Glasgow. He informed his father that if the college would not pay for the apparatus, he would. However, finding that making lists and waiting until winter for faculty approval was an impractical way of purchasing equipment, William gave up the idea of ordering anything else in Paris.[40]

Dr. Thomson couldn't resist a bit of advice to William about his travels: neither "overfatigue, overheat, or overcool" yourself. Thomson's first stop after Paris was Geneva where he stayed with his Cambridge friend Tom Shedden and his wife. Before Shedden settled down to work at the bar, he and his wife were "very wisely, going to see as much that is interesting in Europe as possible." Thomson hoped to travel with the Sheddens for a month, but they had to make a hurried trip to England and Scotland before returning to the Continent for their second winter.[41]

Thomson was delighted with Switzerland and wanted his family to spend their next summer there. Everything including Lake Geneva reminded him of Scotland. He sometimes traveled with a party but most often alone with a guide. He and the guide followed Forbes's directions in order to get the best views of the glaciers. Thomson "saw very well the various peculiarities . . . crevasses, veins, &c." as described in Forbes's book. Through Forbes, Thomson met Alfred Gautier, a professor of astronomy in Geneva. Gautier lived in a country house not far from Geneva with a magnificent view of the lake. The amiable Gautiers made Thomson part of their large family. Conversations in French demonstrated to Thomson that he was not yet fluent, and while in Geneva he had a French lesson every morning.[42]

In his travels through Switzerland, Thomson traveled by mule, carriage, cart, and on foot. While walking one day he met Joule, who was on his honeymoon. Thomson wrote to his father that he was "even more surprised by this accidental meeting than the last," in Oxford. One afternoon he walked eighteen miles and found that the brandy flask and plaid he carried were "most useful" against the effects of the cold.[43] William's tour was memorable. He made a number of trips to the Continent after that, always traveling from place to place in a continual state of motion, taking in the beautiful sights and working on his mathematics at every stop.

From then on, the change in the way Thomson spent his holidays was a marked one. From the quiet months spent at Scottish watersides, in a virtually static appreciation of the country-side, he became more dynamic and was always on the move. This movement was not due to a restless discontent. He enjoyed a passing scene but always managed to keep it moving. The natural beauty of Switzerland was a constant delight; William liked the rough, unfinished variety of scenery here, as well as in Scotland. Movement was an expression of a fundamental part of his personality, and travel gave a shot of energy to the flywheel of his mind. When he stopped for the evening, he caught up with his thoughts and recorded them in his mathematical notebook. For example, at the end of the summer of 1847, an entry in his diary began:

Park House, Maidstone, Kent,   Sep. 21, 1847 (Tuesday)

Despatched [*sic*] today a letter to Liouville, written principally on the top of the Faulhorn (Sep. 12, 1847) but finished last Friday (2 A.M.) and sealed at Bâle. It contained a short paper on distribution of galvanic currents for producing given potentials; also the following, which occurred to me in walking from Sion to Sier on my way to the baths of Leuk, last Wednesday week.[44]

By contrast, time spent with his father and family at their summer places made him feel languid. William wrote the following in his diary: "When I was at home (at Knock) I was principally engaged (when I did any work). . . . " Besides lecturing and examining students at Peterhouse during the summer, he noted that he read a paper before the Cambridge Philosophical Society and wrote a little on a paper in progress, "but besides this, scarcely any thing."[45]

A complement to the characteristic of his mind that enabled motion to stimulate mental activity was his habit of looking back on previous work. At times he recalled having started a problem previously although at the time of writing he could not find it in his diary. Other times he remembered the date he had first worked on the problem and referred to that entry. Often there were marginal and inserted emendations in his diary showing that he often went back to pick up a thread of thought or to make corrections.

For example, in Geneva on 13 August 1847 on the second page of the diary entry, William wrote that a better demonstration was to be found under the entry of 7 October 1847. Later he wrote, "see 22 June 1847." In the margin is a note to himself to see below on 16 August 1848, and the margin of the next page has a correction made in 1871.[46]

As Thomson's mature personality formed, his relations with his family underwent a gradual though perceptible change. A trivial change was that although he continued to visit his family, he no longer spent summers with them. A major change was the fact that no one in his family, including his father, was able to keep up with William's research or enter into conversations with him about his work. Only James, whose engineering background was closer to William's interest, discussed technical problems with him. When it came to William's mathematics, James confessed his inadequacy, but on matters such as Carnot's theory and mechanical problems, James had a different point of view to contribute. There were instances of collaboration between the brothers. In January 1849, William Thomson read one of his brother's papers, "Theoretical Considerations on the Effect of Pressure in Lowering the Freezing Point of Water," to the Royal Society of Edinburgh. The paper began: "Some time ago my brother, Professor William Thomson, pointed out to me a curious conclusion to which he had been led, by reasoning on principles similar to those developed by Carnot, with reference to the motive power of heat." William claimed that if one followed Carnot, water might be converted to ice at zero degrees without the expenditure of any work. The idea struck James as impossible, and in attempting to explain why his brother was wrong he discovered a new hypothesis—later proved experimentally by William—that the freezing point of water is lowered by pressure.[47]

On 12 January 1849, about two years after the death of his son John, Dr. Thomson died. He was sixty-two years old. He was stricken with cholera and died shortly after. At that time, the University of Glasgow was located in the midst of the worst slum in the city, and the periodic epidemics of cholera had been an argument for the removal of the university to another part of the city. When William sent the news of his father's death to David King, his brother-in-law, he wrote: "I could not believe last night at this time that we were to lose him. . . . But God has ruled it for the best, and has tried us with a heavy affliction. . . . It is a terrible and irreparable loss, and a sad void is now left. . . . " There was little else to write. R.L. Ellis's letter of condolence evoked the inexpressible: "My father's death was, of a single event, the greatest grief of my life. . . . More than almost any other loss, the death of one's father disarranges and breaks up the whole system of one's life and of the relations in which the members of his family stand to one another. . . . "[48]

Soon after his father's death, Robert Thomson, the youngest, left Glasgow and emigrated to Australia, never to be seen again and rarely heard from. James Thomson, the oldest son, left Glasgow and settled in London for two years. When his father died, he wrote to a friend: "I felt that I could not live longer with the other members of our family withholding from them my real sentiments [on religion], and I thought that, if I disclosed them, it would probably be necessary for me to live away from home."[49] On the contrary, the family sympathized with James and accepted his conversion to Unitarianism. He moved to Belfast in 1851 and opened a civil-engineering office. In 1857, James Thomson was appointed professor of civil engineering and surveying at Queen's College, Belfast.

William stayed on at his father's college residence, being now the head of the household with his aunt still the housekeeper. The death of his father when Thomson was twenty-four years old should have signaled maturity. However, he had already matured before that time. Although an intensive level of mental activity was suspended, his diary picked up again on 1 February, a few weeks after his father's death.

By 1849, Thomson held firm convictions about the goals of science and a self-assurance to follow where the new ideas took him. In the winter of 1849, at the time of his father's death, he was working on a paper about the mathematical theory of magnetism. When he wrote to Stokes early in February, Thomson referred to an earlier paper, "On a Mechanical Representation of Electric, Magnetic, and Galvanic Forces": "When I wrote the paper [he told Stokes], I had some hope, which I still retain, that a satisfactory physical theory of all those agencies [that is, static and current electricity as well as magnetism] including besides light, is approachable." Thomson apologized for the delay in replying to Stokes: "What has happened since you know, and will readily conceive that my mind was very much turned away from matters which at all ordinary times interest me very much."[50]

During June of 1849, Thomson had transmitted one of his papers, "A Mathematical Theory of Magnetism," to the Royal Society of London. Another step away from the statical view of nature was taken. The opening paragraph rejected the materiality of force and the idea of imponderables. In previous papers on magnetism that were presented by other scientists, Thomson wrote, a hypothesis of two magnetic fluids was adopted and adhered to. "No physical evidence can be adduced in support of such a hypothesis; but on the contrary, recent discoveries, especially in electromagnetism, render it extremely improbable." Although, as Thomson admitted, all the physical effects of magnetism had to be taken into account, such as

Faraday's discovery of the effect of magnetism on polarized light, in order to arrive "at any satisfactory ideas regarding the physical nature of magnetism," he wanted for the present to consider only magnetic attraction and repulsion in presenting a mathematical theory of magnetism.[51] There was not enough experimental data to form a physical theory of magnetism, but it was time to write out the mathematical theory.

Thomson's thinking from 1847 to 1849 reveals the scientific imagination at work. Primarily a mathematical theoretician, he dealt with experimental results in a singular manner. Every one of Faraday's discoveries about magnetism and electricity suggested a physical analogy to Thomson. Not accepting Faraday's explanations, which made use of contiguous particles and curved lines of force, he translated his own physical analogy into mathematical metaphor. The equations represented, without portrayal, the physical forces through their abstract mathematical operations. When Thomson heard Joule's paper, he reacted as he had done to Faraday's explanation and was unwilling to accept Joule's idea that heat was a vibration of matter. Thomson wondered how Joule's results might be expressed in more acceptable mathematical terms. Joule's use of the term "equivalent" provided the link between experimental and theoretical thought. It was the clue to the interrelation of all forms of energy. For Thomson, his thinking on matters of electricity, magnetism, and thermodynamics passed in a gradual metamorphosis from physical analogy to mathematical metaphor so that by a different route from the experimentalists, Faraday and Joule, Thomson arrived at the unifying concept of equivalence.[52]

CHAPTER SEVEN

The Direction of Men's Minds

It is not by discoveries only, and the registration of them by learned societies, that science is advanced. The true seat of science is not in the volume of Transactions, but in the living mind, and the advancement of science consists in the direction of men's minds into a scientific channel; whether this is done by the announcement of a discovery, the assertion of a paradox, the invention of a scientific phrase, or the exposition of a system of doctrine. (James Clerk Maxwell)[1]

The influence of Thomson's original ideas on nineteenth-century science depended on his adherents. His unorthodox use of analogy and mathematical metaphor could not be applied to new problems by others unless they were well enough understood. His theories were sometimes accepted, however, without an appreciation for his method of mathematically interpreting experimental results. In evaluating his role in the advancement of science, two questions arise: Why was he not more successful in winning adherents to his ideas? With so few followers, why did he have such influence on science?

When Thomson wrote "A Mathematical Theory of Magnetism" for the Royal Society of London, he did not expect it to grow into a treatise. Since Thomson was not a member of the Royal Society, he asked Edward Sabine, foreign secretary of the society, to transmit the first part of the first section of the paper. Sabine agreed, and the paper was read in June 1849. In 1844, when Thomson was a student at Cambridge, Archibald Smith had introduced him to the Sabines. Mrs. Sabine made an impression on William. He wrote to his father that she worked closely with her husband and translated "all the very valuable papers of Gauss on Terrestrial Magnetism in Taylor's Scientific Memoirs." He thought the Sabines were "very intellectual and active people." They in turn were impressed with the twenty-year-old student and invited him "to call any time."[2]

The second part of the first section of the paper was read to the Royal Society in June 1850. A section on electromagnetism, originally planned for inclusion, remained in manuscript for twenty-two years and was unpublished until his papers were reprinted.[3] The study of magnetism was a promising path toward understanding the nature of matter and the transmission of force. There were mathematical similarities between the exertion of magnetic force and static electricity. In addition, there were the well-known physical relations between electricity and magnetism discovered by Oersted, Ampère, and Faraday.

A new theory of magnetism was worthy of the Royal Society's attention, Thomson said, because the older mathematical theory was based on the hypothesis of two magnetic fluids, that is, imponderables that were the supposed cause of the attractive and repulsive powers of magnets. No physical evidence had ever been given to support the fluid hypothesis, and recent experiments in electromagnetism rendered the older hypothesis improbable. In writing his paper, Thomson was attempting a complete mathematical theory of magnetism based only on generally known facts.[4] He meant that enough physical effects connected with magnetic force were known to draw a mathematical theory *without making any assumptions* as to the physical nature of magnetism.

Previously, by using Green's idea of potential, Thomson derived the mathematical equations for the action of magnets based on the method of distribution of imaginary magnetic matter. He then rejected these results, simple and convenient as they were, on the grounds that they were too artificial, and although the potential method gave the mathematical expressions for the gross actions of magnets, the method was not precise enough to give an idea of the internal workings of magnets.[5] The gross physical analogy with gravity was inadequate for explaining the more complex phenomena of magnetism. Unlike gravitational bodies, magnets attracted and repelled in addition to being polarized, exerting force at the poles or ends.

In November 1849, Thomson had suggested in his diary that Ampère's idea of imaginary electric currents around molecules of matter was a good starting point for imagining the internal workings of the magnet. Instead of the distribution of imaginary magnetic matter, the question was how might infinitely small electric circuits be distributed throughout a magnet, on the surface and internally, so as to produce the experimentally known effects? One after another, types of arrangement of minute electric wires were tried in search of the simplest form that could be treated mathematically. Thomson finally settled on a complex geometrical arrangement for which he coined the terms "lamellar" and "solenoidal." The equations he derived from this model were similar to those used to describe the motion of a homogeneous incompressible fluid. To Thomson, the similarity was "so obvious that it [was] scarcely necessary to point it out." He did worry that the new terms "lamellar" and "solenoidal" did not convey meaning without some explanation. He knew that scientists such as Stokes would understand after a few words of explanation. The few words to Stokes ran to seven pages.[6]

The ideas, so closely related to hydrodynamics and gravity, were tried out on Stokes. Thomson deferred to few people in his thinking; Stokes was one of the exceptions. In October 1849, when Stokes wrote to Thomson about a new theorem, Thomson's startled reply was: "Is this certainly true? If it is it is most interesting; but at present I am greatly puzzled." He continued:

> I am at present in the middle of a paper on Magnetism the first part of which was communicated last June, and the place where I stopped, in my writing is exactly where considerations like these occur. I have had all the matter ready, and a good deal of it roughly put on paper for a long time, and so I hope soon to get it reduced to publishable form. I am the more anxious on this account to know more of what you have been telling me about. . . .[7]

One of the reasons for confusion, Stokes realized, was his interest in positive quantities only since he was dealing with gravity, and he cautioned Thomson about the difference in results to be expected when dealing with attraction and repulsion. Stokes's letter was long and apologetic, ending with the self-effacing remark that he hoped his imperfect investigations would be of some use to Thomson.[8]

The ideas in "A Mathematical Theory of Magnetism" suggested further research. Thomson felt he had made a significant step forward in consolidating theory:

> However different are the physical circumstances of magnetic and electric polarity, it appears that the positive laws of the phenomena are the same [here he includes a footnote to his paper "On the Elementary Laws of Statical Electricity," published in 1845] and therefore the mathematical theories are identical. Either subject might be taken as an example of a very important branch of physical mathematics, which might be called "A Mathematical Theory of Polar Forces."[9]

Although twenty-two years later, Maxwell still thought Thomson's paper was the best introduction to the theory of magnetism known,[10] the treatise had little effect on the course of nineteenth-century theory. For Thomson, the result of trying to consolidate his thinking on magnetism was to make the ideas readily accessible for practical application. When he received a letter from Archibald Smith asking about the use of iron bars to offset the magnetic effect of iron-hulled ships on a marine compass, Thomson was able, in the midst of his work on the paper, to work out a practical solution based on his mathematical theory.[11] Years later, in Thomson's hands, the same theory was the basis for his successful invention of a mariner's compass.

The mathematical theory of magnetism was valuable for further research and practical application, if only scientists had appreciated its significance. The relatively small impact of such a potentially useful theory illustrates one of Thomson's greatest weaknesses as a scientist—he was not an advocate of his own ideas. As an independent thinker, he was unaware of how separate from others he was in his thinking—to him it was all so obvious. One can only imagine Thomson's reaction when his friend Ludwig Fischer, a Hopkins student and fourth wrangler, now professor of natural philosophy at St. Andrews, complained that he could not understand two of Thomson's papers and wanted Thomson to tell him "what preliminary studies [he] must make before . . . successfully attack[ing] them."[12]

Thomson's distinctiveness as a scientist, which deprived him of intellectual followers, was his genius for maintaining an equilibrium between his restless curiosity and his tenacious hold on such ideas as the belief in the eventual unification of all scientific theory. His mind was an arena in which a dynamical equilibrium existed between adherence to some unchanging principles and an eagerness for new discoveries. There is an analogy between the dynamical equilibrium in his thinking and the way his life was at one and the same time that of a comfortable university don as well as one of the most traveled scientists of his generation.

Thomson's independence of spirit prevented him from being a follower of men, although he was a follower of ideas. He was no man's disciple and followed in no living scientist's footsteps. The ideas that influenced him the most, those of Fourier, Green, and Carnot, were not attached to a personality. Thomson's father was his only attachment in life, and that tie was so perfect that no substitute was possible.

An aspect of his independent attitude was revealed in religious matters; Thomson was an observing but not conforming member of the Church. In these matters, his religious beliefs contrast with those of Faraday, Maxwell, and Stokes, who were greater adherents to a body of tenets. Thomson wrote to Stokes that he was "in the habit of regularly conforming to the Episcopal Church, and not appearing more than once or twice or three times in the course of a session at an Established Church."[13]

Early in 1849, after recovering from the depression caused by his father's death, Thomson considered the matter of his father's successor to the chair of mathematics at Glasgow. Hopkins reminded Thomson to "determine the *principle*," that is, if he wanted to recognize the "scientific principle" in choosing his father's successor, then Stokes was *"unquestionably"* the man; or if he wanted "to recognize the primary importance" and choose one who received his early education at Glasgow, then Blackburn, a man of "acute intelligence and cultivated mind," was his choice.[14]

Understandably, Thomson thought of the advantage of having Stokes in Glasgow to talk mathematics with rather than pass their thoughts through the mail. Thomson's plan was frustrated. Stokes refused to stand because, as his brother advised him, he should not take the religious oath required of all professors at Glasgow unless he planned on becoming a Presbyterian. Stokes refused to do that and was unable to sign the test oath "in a lax sense."[15]

"Your letter which I received this morning has put me quite into a state of agitation," Thomson replied. He did not think the oath was an obstacle for Stokes, and based on his own experience, Thomson was sure the Glasgow professors would sanction Stokes's remaining an Episcopalian and attending the Established Church of Scotland whenever he pleased. As a matter of fact, Thomson had not signed the Confession of Faith and the formula of the Church of Scotland, a declaration of his intention to conform to the worship and discipline of the church. There had been an oversight in his case, he believed, but perhaps it had been a tactful forgetfulness to avoid an altercation with the son of Dr. Thomson, a persistent opponent of the oath. William Thomson had written to Cookson, now master of Peterhouse, "I have long made up my mind, if asked, to refuse to sign, now, or any time in future."[16]

Thomson pleaded with Stokes to reconsider and asked him to "remember that it will be a very serious blow to the interests of this University if an honest member of the Church of England should never be able to be a candidate for any *situation or office* connected with it, however valuable an acquisition he might be; on account of an act of Parliament framed at a period of great political and ecclesiastical excitement. . . . " Stokes was "staggered" by Thomson's vehemence but felt he had no choice but to refuse to stand.[17] Thomson replied:

> I cannot but regret that your determination is finally against what seems to me to be the course that would have been so much better for us, and for Scotland in general than that which you feel to be most satisfactory to your own honourable feelings.[18]

To Thomson, the oath was an unreasonable requirement that free and independent men might circumvent. No such independent spirit motivated Stokes; in his opinion, an oath, no matter how outdated, bound principled men regardless of their own opinions in the matter.

With Stokes's withdrawal, Blackburn's election was a certainty, although Forbes wondered why Thomson had not considered Boole. Blackburn was elected on 13 April 1849. Thomson's intimacy with Blackburn was unlike his relationship with any other of his friends. (Blackburn was the only one of Thomson's correspondents to address him as "Willie.") With Blackburn, he had a personal attachment and some professional dealings. In June 1849, Blackburn married Jemima Wedderburn, an accomplished artist and Maxwell's first cousin. Thomson was his best man. The relationship between the Blackburns and Thomson was a convivial one. On the margin of a page in his diary, Thomson had written: "[We are] in the garden. Blackburn and I smoking pipes and Mrs. B. painting flowers." While touring Italy during the summer of 1851, William stayed with the Blackburns in Florence.[19]

At the university, Thomson and Blackburn expressed a need for better books for student awards. In 1871, they reprinted in a handsome volume Newton's last Latin edition of the *Principia* because they found that all editions of the *Principia* were out of print.[20]

One of Thomson's most valuable contributions to nineteenth-century scientific thought was as a mediator between the experimental and mathematical views of phenomena. In his assessment of Joule's experimental results versus the established mathematical theory of Carnot, he sensed a need for reconciliation. The doctrinaire theorists believed that if Carnot were correct, Joule was wrong. Both could not possibly be right. However, it was not enough to say, as Thomson's brother James did, that "there ought to be some connecting link" between Joule's experimentally determined mechanical equivalent of heat and Carnot's mathematical theory of the motive power of heat.[21] With the vantage point of an independent, Thomson thought that here, as in the dispute over Faraday's lines of force, it was a matter of viewing the same phenomena from two points of view. Reconciliation would provide a better perspective of the way heat produced mechanical motion.

In February 1851, Thomson began the draft of a paper on heat and included the notation "state axiom."[22] He planned to start his paper with the unequivocal statement that heat was the result of the motion of particles and that it was not a substance. Thomson was prepared to accept Joule's controversial theory of heat. However, his role as mediator caused him some trouble. The drafts of the discarded versions of the paper reveal some of the struggles involved in his mental transition.

In his second try at the draft, Thomson claimed that when he had written his account of Carnot's theory two years before, he "felt very great doubt as to the fundamental axiom," although he was not then prepared to accept Joule's contrary axiom. On 1 March 1851, the draft began with Thomson's original objection to Joule. While Carnot's theory stated that when heat was conducted away from the engine, the mechanical effect was *"absolutely lost,"* in the dynamical theory, the heat conducted away could not be accounted for in any other way than to insist that *"it is not lost. . . ."* Thomson believed the solution to the dilemma was in the "fact," not yet demonstrated, that the work was *"lost to man* irrecoverably; but not lost in the material world." Energy was destructible only by an act of the Creator. "Everything in the material world is progressive. The material world could not come back to any previous state without a violation of the laws which have been manifested to man; that is without a creative act or an act possessing similar power."[23]

In the material world, the ways of God were predictable in contradistinction to the unpredictability that existed in the "moral world," or the world of the mind. The draft read:

> I believe the tendency in the material world is for motion to become diffused, and that as a whole the reverse of concentration is gradually going on—I believe that no physical action can ever restore the heat emitted from the sun, and that this source is not inexhaustible; also that the motions of the earth and other planets are losing vis viva which is converted into heat; and that although some vis viva may be restored for instance to the earth by heat received from the sun, or by other means, that the loss cannot be *precisely* compensated and I think it probable that it is undercompensated.[24]

In answer to those scientists who thought that the losses were compensated, Thomson supported the biblical opinion, "The earth shall wax old &c."

Mechanical effect escapes not only from agencies immediately controlled by man, but from all parts of the material world, in the shape of heat, and escapes *irrecoverably*, though without *loss of vis viva*. [25]

These thoughts were Thomson's first expression on the implications of a scientific theory, marking another departure for him. The apparently conflicting theories of Joule and Carnot impelled Thomson to worry about cosmological and theological questions. It was significant that theology was not discussed in the published paper. Thomson had definite opinions about the separation of scientific thought from theology, although he believed God's role was implicit. The suggestion of the cosmological implications of the second law of thermodynamics did appear in his published paper, but by then the scientific theory was so far along toward acceptance that the issue was not raised at that time.

Further work on the paper, made on 10 March, began with a mathematical treatment of a perfect thermodynamic engine. By this time, Thomson was worried about his deadline and admonished himself to begin in earnest. The energy of molecular motions was now taken as the essence of the dynamical theory of heat, and he viewed the communication of heat as being due to the transmission of motion between contiguous molecules. [26] This thought suggested a similarity to Faraday's theory that electric force was transmitted through the action of contiguous molecules.

Finally, on 13 March, Thomson started the version of his paper in the form that is very close to the one published. He wrote at the top of the page, "On the Dynamical Theory of Heat; with numerical Results deduced from Mr. Joule's Equivalent of a Thermal Unit, and M. Regnault's observations on Steam," the title as published. The final version of the paper had no support for the dynamical theory of heat other than historical. The theory had first been proposed by Sir Humphry Davy in 1799, and this speculation was confirmed by recent experiments performed by Joule and Mayer. This evidence was the only support for Thomson's statement: "Considering it as thus established, that heat is not a substance, but a dynamical form of mechanical effect, we perceive that there must be an equivalence between mechanical work and heat, as between cause and effect." [27]

One of the objectives of this paper was to determine what modifications needed to be made on Carnot's conclusions in light of the dynamical theory. Instead of being a conflict, Thomson declared, "the whole theory of the motive power of heat is founded on the two following propositions, due respectively to Joule, and to Carnot and

Clausius."[28] The reconciliation of Carnot's theory with Joule's experimental results was contained in a statement made by Thomson:

> The demonstration of the second proposition [Carnot's] is founded on the following axiom:
> *It is impossible, by means of inanimate material agency, to derive mechanical effect from any portion of matter by cooling it below the temperature of the coldest of the surrounding objects.*[29]

This statement was one form of what would later be called "the second law of thermodynamics." It has been judged one of the most important breakthroughs in nineteenth-century science. The second law, in addition to the first, which deals with the conservation of energy, became the basis for the study of heat in its many manifestations in nature and its conversion to useful energy. The implications for scientific theory were manifold, and in technology it became the fundamental axiom for the design of all engines employing heat as well as in cooling and refrigerating systems.

Ironically, or perhaps typically, in the history of science what had recently been a controversial theory now became the center of a bitter dispute over priority. Thomson's attitude toward priority reflected his broad approach to his work: "Questions of personal priority, however interesting they may be to the persons concerned, sink into insignificance in the prospect of any gain of deeper insight into the secrets of nature." In this article, Thomson readily relinquished claim to priority. He was even more forthright in the early draft, "I may be allowed to state here that I lay no claims to discovery in this theory. . . . "[30] In an article published ten months before Thomson's article, Clausius had made a statement on what amounted to the same idea as Thomson's. Although he had been anticipated, Thomson insisted that his own work was independent since he had not heard of Clausius's paper until after his own was completed. In one way or another, Julius Mayer, Joule, Clausius, Rankine, Liebig, Helmholtz, and Thomson had some claim to credit for either the first or second law of thermodynamics.[31] Thomson's most important contribution was the usefulness of the way he made his statement and the way he followed the implications of the law. Even if it be granted that Mayer had the first adumbration of the law, it was Thomson's broadening of the concept that made the law so important to science and the reason why so many scientists argued over its priority.

As Thomson pointed out, Carnot's theory "expresses truly the greatest effect which can possibly be obtained in the circumstances;

although it is in reality only an infinitely small fraction of the whole mechanical equivalent of the heat supplied. . . ." Engineers were interested in the "infinitely small fraction," and scientists gave attention to the gross quantity, which Thomson said was "irrecoverably lost to man, and therefore 'wasted,' although not *annihilated*." The modification of Carnot's theory by the dynamical theory of heat did not alter any conclusions in his "Account of Carnot's Theory," and the idea developed by his brother James in "Theoretical Considerations on the Effect of Pressure in Lowering the Freezing Point of Water" still pertained, Thomson said.[32]

As shown in this paper, Thomson's interest was the practical application of the mathematical theory (Carnot's) by means of the experimental results of Joule and Regnault. Not only did Joule's findings require a basic modification of Carnot's theory to include the dynamical theory, but Joule's calculated results along with Regnault's contained the data for calculating the efficiency of specific engines and enabling engineers to design steam engines better before they were constructed. The length of the paper was considerably extended in making just those correlations between the theory and the data:

> Thus we see that, although the full equivalent of mechanical effect cannot be obtained even by means of a perfect engine, yet when the actual source of heat is at a high enough temperature above the surrounding objects, we may get more and more nearly the whole of the admitted heat converted into mechanical effect, by simply increasing the effective range of temperature in the engine.[33]

Joule was right. No heat was lost in the conversion of heat to mechanical effect, but the largest part of the potential mechanical effect contained in a source of heat, say steam, was unavailable. Since it was impossible to reduce the condenser of a steam engine to absolute zero, it was impossible to convert all the heat in any source to mechanical effect. Heat was irrecoverably lost to man, although still in the steam at the exhaust of the steam engine and, therefore, not annihilated. The second law, as stated by Thomson, said that mechanical effect could not be obtained from the heat in matter if you cooled that matter below the temperature of surrounding objects, and the coldest of those surrounding objects was considerably above absolute zero.

In a footnote to his paper, Thomson wrote that he was now prepared to accept Joule's contention that heat was put out of existence in the production of mechanical effect, and this was something Thomson was unwilling to admit in 1847. But Thomson still insisted that no *experimental* evidence proved this contention. In no case of the "production of mechanical effect from purely thermal agency, has the ceasing to exist of an equivalent quantity of heat been demonstrated otherwise than theoretically."[34] The skepticism of the mathematical theorist prepared the way for a long period of collaboration with an experimentalist.

The first part of the dynamical heat paper elicited a letter from Joule who had comments and questions about the theory. This exchange of letters led to the beginning of a period of collaboration from 1852 to 1856. Much of the cooperation was done through the mail with an occasional trip by Thomson to Manchester. These were flying trips for a day or so. Whenever Thomson stayed in Manchester, Joule arranged for a quiet room where Thomson could write while he attended to brewery business.[35] Joule rarely traveled to Glasgow.

Joule designed, often constructed, and usually conducted the experiments by himself. They received a grant from the Royal Society to construct some of the equipment. It was Thomson's responsibility to test the data and see if it conformed to theory. Joule directed Thomson, "Now I want you to consider how it is that the results should seem to point out that the specific heats are the same for equal volumes contrary to Regnault's Experiments."[36] During the period of their joint effort, their association assumed a pattern; the following is a typical letter from Joule:

Southport, August 13, [18]52

My dear Thomson,

   I have taken the following plan in carrying out the experiments on the temperature of rushing air. A brass pipe was soldered to the lead coil and a piece of calf skin leather was tightly bound over the open end. Over this a piece of vulcanized india rubber piping was fastened. In the experiments the compressed air rushed through the pores of the leather against the bulb of a very small thermometer placed in the india rubber tubing, the bulb being in immediate contact with the leather. The pressure was measured as before by a mercurial gauge. The following results were obtained:

   [Here Joule inserted the results of two series of readings in which he varied the pressure and kept the temperature constant, first at 61 degrees Fahrenheit and then at 160 degrees.]

You will observe that with the high temperature, the air is cooled in rushing through the porous leather, but not to more than about half the extent that was observed with the lower temperature. Probably with a temperature of about 250° Fahr the cooling effect would disappear altogether, and beyond that a heating effect be observed.

Are not these results nearly what you anticipate from theory?

I am not quite sure whether I shall be able to go on with higher temperatures, but will endeavour to do so if you think it desirable.

Believe me ever yours

James P. Joule

P.S. Direct as before to Acton Sq—I have been much hurried of late and am here only for 2 or 3 days for the benefit of sea air—.[37]

Joule asked for and received suggestions for further experimentation. Some of Thomson's letters suggested new series of experiments.

I now understand I think completely the principles of all the tests you apply in your paper on the "changes of temperature &c of air" and am most firmly convinced of their correctness, but I still think it is very much to be desired that as accurate a test as possible should be applied both to air at ordinary atmospheric temperatures and at various temperatures between as high and as low limits as possible. . . . I think no [sic] so limited series of experiments could do more to advance the theory of heat from the present state of science than a series to determine the relation between the mechanical effect and the heat in the compression of air at different constant temperatures. I do hope you will be induced to enter upon the subject again.[38]

The article reporting their results was titled "On the Thermal Effects of Fluids in Motion" and filled 122 printed pages.[39] The data were used for a more accurate calculation of Carnot's function; the accuracy measured up to Joule's standards with such readings as intervals of temperature of 0.011 degree centigrade. Thomson and Joule saw the usefulness of the experiments in calculating the efficiency of steam and air engines as well as advancing the theory of heat and thermoelectricity. The work was a total confirmation of Joule's original theory on the conservation of energy and a justification for his mechanical equivalent of heat. They amassed enough evidence so that both experimentalists and theorists had to accept the validity of Joule's early work. Their collaboration served its purpose by testing experimentally and confirming theoretically an idea, but that was the limit of their endeavor. Thomson and Joule had discovered nothing new as Joule had done when he worked alone, nor had they extended theory into new areas as Thomson had done alone. When they worked together, they amassed a great deal of useful data. However, science was advanced much further as a consequence of their disagreement over the dynamical theory of heat.

In June 1851, just before his twenty-seventh birthday, William Thomson was elected a fellow of the Royal Society of London. The honor was greater because of his relative youth. Stokes was made a fellow the same year; he was thirty-one. Thomson's contribution to science was measurable in the fifty-four papers he had written,[40] and recognition of his work was gradually spreading.

A large proportion of his published papers had appeared in the *Cambridge and Dublin Mathematical Journal* whose readership and contributors included some of the most promising scientists of the day. Nevertheless, it was still not a major scientific journal.

Thomson was becoming weary of his job as editor of the *Cambridge and Dublin Mathematical Journal* and asked Stokes to succeed him. Thomson had found himself feeling impatient with the papers submitted and told Stokes that he would be happy to publish his paper "as I am very desirous of getting such papers on physical subjects sometimes in place of the endless Algebra and combinations which so abound."[41] After Stokes declined the editorship, Thomson finally found N. M. Ferrers, fellow of Gonville and Caius College in Cambridge.

A few of Thomson's more abstruse papers appeared in Liouville's *Journal de mathématiques.* Accounts of his papers delivered at the annual meetings of the British Association[42] were published in the association's *Reports.* Some of his papers were published in the more widely read *Philosophical Magazine.* Thomson's paper on the absolute temperature scale appeared in the *Proceedings* of the Glasgow Philosophical Society, of which he had been a member since 1846. In 1847, he was made a fellow of the Edinburgh Royal Society, where his papers on Carnot's theory and the dynamical theory of heat were read. The mathematical theory of magnetism was transmitted to the Royal Society of London for him by Colonel Sabine since Thomson was not yet a member.

Although recognition by the Royal Society was warranted, it must be admitted that in 1851 Thomson's scientific research was contained within a small sphere of influence. The public of that day had not heard of him, and he was not yet of historical significance in the advancement of science. Had he died before the age of twenty-seven, or had he stopped making original contributions to science and technology, his name would be no better known today than many other scientific worthies of his day, such as Ellis or Gregory.

Thomson's most important contributions to science, contributions that had possibilities for changing the direction of scientific thought, had not been taken up by an interpreter who was capable of realizing their potential. Conceivably, these ideas might have remained dormant and discovered a generation later as still another example of a man whose ideas were not appreciated fully during his lifetime. But Thomson did find his interpreter during his lifetime, or rather his interpreter discovered him; he was James Clerk Maxwell.

The backgrounds of the two men were remarkably similar. Maxwell's mother died when he was eight; Thomson's when he was six. They both went to Cambridge after attending Scottish universities— Maxwell went to Edinburgh and Thomson, Glasgow. They were both Hopkins's students and second wranglers, but Maxwell did not match Thomson's record as first Smith's prizeman; while Thomson beat Parkinson handily, Maxwell was bracketed as equal to Routh. The two men probably met at Hugh Blackburn's wedding in 1849. Earlier that year, Maxwell had consulted Blackburn about Cambridge,[43] and heeding Blackburn's advice, Maxwell went to Cambridge. He matriculated at Peterhouse, Thomson's college, and migrated to Trinity, Blackburn's college.

In August 1850, Thomson made Maxwell a party to his desire to correct John Tyndall's experiments on crystals. Thomson was troubled by what he thought were experimental errors reported in an article coauthored by Tyndall. He wrote a letter to Stokes criticizing Tyndall's work and subsequently asked Stokes's help in determining the exact basis for Tyndall's errors. When Thomson met Tyndall at the meeting of the British Association in Edinburgh, in 1850, they discussed Thomson's objections to the article. Thomson now felt more beneficent and sought to resolve the controversy.[44] Then he discovered that some experiments by young Clerk Maxwell raised the possibility of doing just that and urged Maxwell to continue his investigations. Soon after Thomson returned to his summer place at Row, he wrote Tyndall an eighteen-page letter laying the path to settling the affair.[45] It was a case in which a scientific issue briefly aroused Thomson, but his good nature prevailed.

Early in 1854, just after finishing his B.A., Maxwell sent Thomson a letter and asked him to advise a course of study for learning about electricity; a group at Cambridge wanted to prepare themselves for research in the subject. In November he wrote to Thomson as, he said, "an electrical freshman." The complex subject was clearing up with the help of a few simple ideas, especially Thomson's analogy with heat. He thought about representing Faraday's magnetic lines of force, which Thomson mentioned, by a geometrical construct. Having described his state of knowledge at the moment, Maxwell wanted to know what else Thomson had published and what else he recommended for reading. Thomson suggested Weber's work, which Maxwell did not like.[46] In their correspondence, Thomson's role was mentor, and Maxwell, that precocious student, chatted on about numerous subjects—his experiments in the theory of colors, hydrodynamics, and dynamics—as well as asking for suggestions on teaching natural philosophy.

By the late summer of 1855, Maxwell felt certain enough about his understanding of electrical matters—due in great part, he said, to Thomson's tutoring by mail and his published papers—to want to write something on the subject himself. He wanted to write a mathematical theory of the attraction and induction of electric currents as well as of attraction by electrified bodies, providing Thomson did not already have "the whole draught of the thing lying in loose papers and neglected only till [he had] worked out Heat or got a little spare time." Then Maxwell made a specific list of his indebtedness: Thomson knew of Faraday's ideas about lines of force, Ampère's current laws, and "of course you must have wished at least to understand Ampère in Faraday's sense." Thomson, Maxwell knew, was well acquainted with Green's essay and the concept of potential, and part of Thomson's speculations were already published in the analogy of electricity with incompressible elastic solids. Thomson's paper on magnetism had stated and applied Ampère's idea about circulating current to solenoidal distribution of magnetism. Thus Maxwell urged Thomson:

> As there can be no doubt that you have the mathematical part of the theory in your desk all that you have to do is to explain your results with reference to electricity. I think that if you were to do so publicly it would introduce a new set of electrical notions into circulation and save much useless speculation.[47]

The answer must have been no. Thomson did not have a draft of such a paper in his desk. Nor is it likely that he had in mind doing what Maxwell contemplated. In December 1855 and the following February, Maxwell delivered a two-part paper to the Cambridge Philosophical Society, "On Faraday's Lines of Force." By unifying ideas about electricity into a single theory, Maxwell proposed to use a physical analogy "to obtain physical ideas without adopting a physical theory. . . . " He planned to do something similar to Thomson's analogy with heat:

By the method which I adopt, I hope to render it evident that I am not attempting to establish any physical theory of a science in which I have hardly made a single experiment, and that the limit of my design is to shew how, by a strict application of the ideas and methods of Faraday, the connexion of the very different orders of phenomena which he has discovered may be clearly placed before the mathematical mind.[48]

To obtain a purely geometrical representation of Faraday's lines of force, Maxwell used an imaginary incompressible fluid flowing in tubes. The velocity and direction of the fluid represented the intensity and direction of the lines. The fluid was in no way to be construed as ordinary or even as hypothetical. It was merely a collection of purely imaginary properties used for writing theorems in pure mathematics in a way that was more intelligible to many people and easily applicable to physical problems.[49]

By giving six mathematical laws expressing Faraday's mode of thought, Maxwell insisted that he was not giving anything resembling a physical theory and, in fact, the chief merit of his paper, he said, was its use as a temporary instrument of research and did not "*account for* anything." What was the use of imagining such purely hypothetical ideas? Maxwell's answer was reminiscent of Thomson's words written ten years before. Maxwell thought it was good to have two ways of looking at a subject.[50]

In 1861 Maxwell published a paper, "On Physical Lines of Force." He said that his previous paper, "On Faraday's Lines of Force," was for the geometer, but this one was written from a mechanical point of view. The object was to determine what the action of Faraday's medium had to be in order to account for the known phenomena. Maxwell wrote to Thomson that his idea of a dynamical theory of magnetism was derived from an 1856 paper by Thomson, "Dynamical Illustrations of the Magnetic and the Helicoidal Rotatory Effects of Transparent Bodies on Polarized Light."[51] "On Physical Lines of Force" was Maxwell's fullest extension of an electromagnetic model. Although it appalled most mathematicians at that time, years later it was being referred to by engineers as a reasonable way to imagine electromagnetic phenomena. Later, in reviewing the reprint of Thomson's papers on electrostatics and magnetism, Maxwell gave the exact quotation from Thomson that suggested the idea of a dynamical medium:

> The explanation of all phenomena of electro-magnetic attraction or repulsion, and of electro-magnetic induction, is to be looked for simply in the inertia and pressure of the matter of which the motions constitute heat. Whether this matter is or is not electricity, whether it is a continuous

fluid interpermeating the spaces between molecular nuclei, or is itself
molecularly grouped; or whether all matter is continuous, and molecular
heterogeneousness consists in finite vortical or other relative motions of
contiguous parts of a body; it is impossible to decide, and perhaps in vain
to speculate, in the present state of science. [52]

Maxwell noted that Thomson's remarks were made in 1856 and
also mentioned that in 1861 and 1862 his own theory of molecular
vortices applied to magnetism and electricity appeared, "which may
be considered as a development of Thomson's idea in a form which,
though rough and clumsy compared with the realities of nature,
may have served its turn as a provisional hypothesis."[53]

In December 1864, Maxwell's "Dynamical Theory of the Electromagnetic Field" was read to the Royal Society. This landmark of
nineteenth-century science was the capstone of the mathematical
study of electricity which Thomson had begun in 1842 with the
publication of his "Uniform Motion of Heat in Homogeneous Solid
Bodies, and its Connexion with the Mathematical Theory of Electricity." That is what Maxwell meant when he said that most of what he
knew on the subject he owed to Thomson.[54]

Maxwell's electromagnetic theory, in spite of his earlier use of
physical analogies and physical models, is a pure mathematical theory. He made no attempt to conceptualize the phenomena physically. The theory is a super mathematical one that by means of
mathematical metaphor not only explains the then known varied
effects of electricity and magnetism but also as interpreted by successors such as Heaviside, Hertz, and Fitzgerald, has been found to
extend to the present time so that Maxwell's equations mathematically describe all known electromagnetic phenomena.[55]

As pure mathematics, Maxwell's theory has been difficult for experimentalists to interpret. It was more than twenty years later that
Heinrich Hertz tested Maxwell's theory of the electromagnetic nature of light. Hertz's contribution was to interpret physically Maxwell's equations. It was one thing for Maxwell to say that light waves
and electromagnetic waves were one and the same, mathematically,
but what were the tests of the idea? What experiments needed to be
done? The equations are abstractions that include all phenomena
but contain no suggestions for the experimenter to test. Hertz's experimental proofs were a crucial step in the advancement of the
theory, and they were by no means obvious. The wireless age then
has its origins in the mathematical theory of Maxwell and the experimental proof by Hertz.

In contrast, Thomson's work on the dynamical theory of heat

resulted in mathematical equations which, when the constants were determined by experiment (Regnault's and the Joule-Thomson experiments), were a basis for calculating steam engine efficiency and for the design of steam engines. Thomson's application of Fourier's mathematical theory of heat when supplied with observations of terrestrial temperatures became the basis of his later work on the age of the earth. "The Essay on the Figure of the Earth" and his mathematical work on gravitational attraction led Thomson to his work in inventing tide measuring machines. In Thomson's hands, the mathematical theory of terrestrial magnetism became the basis for the design of a mariner's compass that compensated for use on iron ships. The mathematical analogy between heat and electricity was the theoretical basis for Thomson's practical solution to the problem of transmission through submarine telegraph cables.

Was Maxwell's theory still another example of Thomson being anticipated? Did he indeed have the draft of the paper in his desk? No, because no matter how much Maxwell owed to Thomson's original ideas, Maxwell's work was a departure from Thomson's way of thinking.

The two men's words contrast their objectives. Maxwell wrote: "When I had translated what I considered to be Faraday's ideas into a mathematical form, I found that in general the results of the two methods coincided. . . . [56] Thomson wrote that his papers "constitute a full theory of the characteristics of lines of force, which have been so admirably investigated experimentally by Faraday, and complete the analogy with the theory of the conduction of heat, of which such terms as 'conducting power for lines of force' . . . involve the idea."[57] Maxwell did "translate" Faraday into mathematical form, but then an experimental interpretation was necessary. The breadth of Thomson's germinal idea was such that on the one hand it became the basis for Maxwell's theory, and on the other hand it formed the theoretical basis for the Atlantic cable.

CHAPTER EIGHT

Science and Technology

If natural science ever furnished a theme for a poet, it is to be found in this achievement.[1]

In September 1852, when he was twenty-eight years old, William Thomson married Margaret Crum, his second cousin. Her father, Walter Crum, was head of a calico-printing firm, and the family lived at the Rouken, Thornliebank, near Glasgow. The two had known each other since childhood. Margaret was often at the Thomson house visiting William's sisters. In writing to Mrs. Elizabeth Thomson King, her future sister-in-law, Margaret foretold an untroubled marriage:

> We have one interest in common that can never fail, and as I told Mrs. Gall [Aunt Agnes], I feel that in William's love for his sisters and her, lies my best security for the continuation to me of those feelings on which the happiness of my life must now depend.[2]

Thomson wrote to Stokes of his impending marriage: "I cannot describe her exactly to you, but I am sure that is unnecessary to ensure your good wishes at present, and when you come down to us in Scotland, I am sure you will be glad to make her acquaintance."[3]

In 1857, at the age of thirty-eight, Stokes married Mary Robinson, daughter of the Reverend Thomas Romney Robinson, astronomer of the Armagh Observatory. Before her marriage Mary had strong misgivings about life with a man who was engrossed in scientific research. Her hesitation came in spite of the fact, or perhaps because, her father was a hard-working scientist. Their engagement reached a crisis when Stokes decided to write a thoroughly honest and thoughtful disclosure of his attitude toward his scientific work and his feelings about marriage. More than anything else, Stokes wanted a wife, children, and the affection that went with family life. Stokes understood his fiancée's trepidation:

I know or fear that I am, or perhaps may almost say used to be, cold. With me, in my life of isolation and abstraction, affection has been in the condition of a virtue requiring cultivation. . . .

Up to the moment of the wedding, he feared she might "even now draw back, nor heed though I should go to the grave a thinking machine unenlivened and uncheered and unwarmed by the happiness of domestic affection."[4]

In May 1853, Thomson and his bride of eight months spent their holiday touring the Mediterranean. On the trip, Mrs. Thomson's health broke down under the exertion of travel. Still sick when they returned home, Margaret Thomson was taken to Edinburgh for surgery, where she remained through the winter for convalescence. By summer, very little progress had been made toward restoring her health. She looked better but was unable to walk any distance. Thomson carried his wife up and down the stairs at home and often had to carry her from room to room. A walk halfway around the garden forced her to bed for several days. "By avoiding all such exertion she keeps tolerably free from pain," Thomson wrote to his sister Elizabeth, "and has much the appearance of good health."[5] No one said, if they knew, what the illness was.

The gravity of Mrs. Thomson's illness cast a pall on Thomson's life and work. He vacillated between hope and despair—hope that the small signs of improvement meant recovery, and despair when the inevitable relapse came. Many letters from friends contained questions about the state of Mrs. Thomson's health, but J. D. Forbes was especially solicitous. He urged patience and hope, gave sympathy, and advised on the efficacy of various cures.[6] At times, the periods of despair affected the flow of Thomson's letters.

On 20 February 1854, Thomson wrote to Stokes: "It is a long long time since I have either seen you or heard from you, and I want you to write me about yourself and what you have been doing since ever so long." Stokes replied: "It is certainly a long time since we have had any communication with each other, hardly since your marriage."[7] It was a good sign that Thomson reopened his correspondence with Stokes. Writing to Stokes was Thomson's way of unburdening his mind; the letters concentrated on scientific ideas. One evening, Thomson could not sleep until he wrote to Stokes about a conversation he had with Forbes: "I wished to ease my mind before going to bed. . . ."[8]

There was no improvement in Margaret Thomson's health for seventeen years. Many different cures were tried—the waters of Krueznach being those particularly recommended. Different physicians were consulted. At last, Thomson lost all hope of finding a cure and only looked for an alleviation of his wife's pain.

On a visit to Kreuznach in the summer of 1855, Thomson arranged to meet Hermann von Helmholtz, who was then professor of physiology at Königsberg. In 1871, he was appointed professor of natural philosophy at Berlin. Born in 1821, he was three years older than Thomson. He had gained a reputation by the publication of *On the Conservation of Energy* in 1847. In a letter to his wife, Helmholtz described his first meeting with Thomson:

> I expected to find the man, who is one of the first mathematical physicists of Europe, somewhat older than myself, and was not a little astonished when a very juvenile and exceedingly fair youth, who looked quite girlish, came forward. He had taken a room for me close by, and made me fetch my things from the hotel, and put up there. He is at Kreuznach for his wife's health. She appeared for a short time in the evening, and is a charming and intellectual lady, but in very bad health. He far exceeds all the great men of science with whom I have made personal acquaintance, in intelligence and lucidity and mobility of thought, so that I felt quite wooden beside him sometimes.[9]

Thomson resumed his entries in his mathematical diary early in 1855. Finally, he was able to adjust to a way of life with an invalid wife without any slackening of his work. Indeed, this was a period when the usually impatient man managed his most prolonged projects. He was indefatigable: his concern for his wife was constant and thoughtful, while his research continued along the lines long since started. Whatever personal and social disruption he endured neither depressed, wore him down, nor distracted him from the resolve to pursue his scientific ideas. The continuance of active thinking in Thomson's mathematical diary contrasted with the inactivity forced upon him through his personal calamity.

The Atlantic cable was a physical liberation for Thomson. He willingly went on the cable-laying expeditions and found in the activity connected with technology the kind of movement that the summer tours of the Continent had supplied in the past. He was a dutiful husband. Whenever he was able, he went to the watering places, which he found dull. During the university session or when he was away, Mrs. Thomson went for the rest cures without him. He did not seek to escape the confinement of being an invalid's companion, but the demands of the cable were met. He never excused himself on the justifiable grounds that his wife needed him. Something was needed as an outlet for this dynamic man after six months of university lecturing, and the Atlantic cable was a godsend.

The Atlantic cable was the most impressive achievement of electrical technology of the age. To William Thomson, it demonstrated how the principles of science were useful in technology. During one of the celebrations after the successful completion of the cable, Thomson was called upon to respond to the toast "Science as applied to Telegraphy." His remarks carefully separated the connection suggested by the toast. Instead, he spoke of the need for scientists to do their research "out of a pure love of knowledge, or from an abstract desire to become acquainted with the laws of nature. . . . " Furthermore, scientists had no greater reward, he said, than to know that their ideas were the means of conferring a practical service to mankind. Abstract science, he believed, was not essential to the advancement of technology. But science accelerated technology by eliminating the need for cut-and-try methods.[10]

The Atlantic Telegraph Company was founded in December 1856, originating as an agreement between Cyrus W. Field, John Watkins Brett, and Charles Tilston Bright. Field was an American businessman with no previous cable experience who had retired at thirty-five. When the object of the enterprise was to connect the eastern shore of Newfoundland with New York City in order to cut several days off communication between Britain and the United States, he became part of the scheme. Before the plan had gone very far, he was captured by the idea of placing a cable across the Atlantic.

Good sense brought Field to Britain where he contacted some people with cable experience. Brett, with his brother Jacob, had successfully laid submarine cables to France and Belgium. Bright had become chief engineer of a British telegraph company when he was twenty and had supervised the laying of the cable to Ireland. When Bright was appointed engineer in chief for the Atlantic Telegraph Company in 1856, he was twenty-four. The primary agreement provided a special compensation for the four original projectors: Field, Bright, Brett, and E. O. W. Whitehouse. Formerly a physician, Whitehouse had given up his medical practice to devote full time to submarine telegraphy. Like so many other telegraph inventors (Morse, Cooke, and Edison), he was self-taught. Whitehouse was named chief electrician for the company. Of the four original projectors, Field knew virtually nothing of cables but was knowledgeable in arranging financing. Brett and Bright both had successful experiences with cables much shorter than the projected Atlantic cable. Whitehouse's reputation as a practical electrical technician was based on papers he gave to the British Association beginning in 1854.

The idea of an Atlantic cable contained the romance of a caravan to the East and promised to be more profitable. The *Times* reported regularly the handsome and uninterrupted dividends paid by cable companies. Submarine cables had piqued the British imagination and the *Times*'s vision became global:

> India soon, and Australia, will be brought within the current of the electric stream, which will carry the advices [*sic*] and the commands of England with the speed of lightning through every portion of the empire.[11]

Financing an undertaking as large as the Atlantic cable was a major task, and Field was able to raise the largest portion of the capital in London, Liverpool, Manchester, and Glasgow. Eighteen of the directors were British, nine were Americans, and three were from Canada. William Thomson was elected to the board of directors from Glasgow in December 1856.

The British government promised Field a subsidy of £14,000 a year when the line was completed. The American Congress, though more reluctant, eventually granted an equal amount of support. In addition, Field procured the aid of both countries in surveys of the ocean floor and each supplied ships for the cable-laying expeditions.

The electrical problems of the projected Atlantic cable received scant attention. For this extraordinarily long cable (over 1,900 miles) manipulating a wire of that length and bulk presented the most serious difficulties. The constant concern was that strain on the cable during the laying operation might lead to breaking the cable. Engineers who attended sessions of the Institution of Civil Engineers, the only professional organization of engineers at the time, devoted discussions about submarine telegraphs to the problems of the laying-out process. During one session, two papers on submarine cables were discussed over four consecutive meetings, and the report of the discussion covered 148 pages of the *Proceedings*. The third paper at that session dealing with electrical properties of cables was read in abstract because too much time had already been taken up with cable topics.[12]

The oceanographic feature of laying a cable across the North Atlantic appeared to be an extension of previous submarine cable projects. Was there a route without unreasonable depths? In 1854, Lieutenant M. F. Maury, head of the Naval Observatory in Washington, interpreted an oceanographic survey and discovered a plateau on the bottom of the North Atlantic that was ideally suited for the location of submarine cables. It was reasonably shallow, 2,000 fathoms at the deepest, and a cable lying on the plateau would be out of reach of icebergs and ships' anchors. Maury realized he had an answer to one difficulty only. He wrote in his report that he did not "pretend to consider the question as to the possibility of finding a time calm enough, the sea smooth enough, a wire long enough, a ship big enough, to lay a coil of wire 1600 miles in length. . . . "[13]

With the general interest raised over the practicality of an Atlantic cable, all aspects were considered. When it was reported that the submarine cable from Harwich to The Hague affected signals transmitted over it by slowing them down and making the received pulses less sharply defined, the question arose whether these difficulties would be an obstruction to working a cable over 1,000 miles long. Faraday conducted experiments on submerged cables of 1,500 miles in length and found the delay in receiving the signals appreciable, two seconds, but not so great as to impede the operation of the projected Atlantic cable.[14] Faraday likened the effect to that of a Leyden jar when being charged, and he calculated the dimensions of an equivalent Leyden jar which would produce the effects he observed.

What was the Atlantic cable? Was it an immense laboratory experiment which demonstrated the Leyden-jar effect of long submarine cables, as Faraday saw it? Was it the possibility of tying industrialized Great Britain with the economic potential of the United States that was just beginning to be realized? Was it a theme for poets, as the *Times* thought of it, with all the romance of instant communication across a vast natural barrier? Was it the promise of great profits for daring entrepreneurs such as Field and the group of investors who readily subscribed the capital? Or was it the naval feat of conquering the Atlantic Ocean, as Maury saw it?

In the excitement surrounding the planning of the Atlantic cable, Thomson found himself in a gradually changing role. He had always seen scientific theory as a guide to technology. Theory was a means of improving plans and designs as well as a means of avoiding pitfalls when technology ventured into untried paths. Thomson showed no direct interest in the application of theory, such as the use of steam power for industrialization and as part of the railroad-building craze. His thermodynamic research stopped short of the machinery itself. He paid no attention to the boom in telegraph expansion or to the multiplying use of submarine telegraphs. He remained interested only in the abstract mathematical theory of electricity.

All that began to change with the letter Stokes wrote to him on 16 October 1854: "Am I right in attributing the finiteness, and even (for such experiments) considerable magnitude, of the time concerned in the phenomena described by Faraday . . . relative to the charging &c. of a long (100 miles) telegraph wire coated with gutta percha and immersed in water to the following two causes?" Stokes's closing sentence was, "I should be obliged to you to write me a line to say whether you think this the true view."[15]

Thomson responded in two long letters. One, the letter of 28 October 1854, was an application of his earlier theory; the other, the letter of 30 October, was more specifically directed to the Atlantic cable problems he foresaw. Since he thought these ideas would be of interest to scientists, Thomson sent his letters to the Royal Society for publication saying they might "serve to indicate some important practical applications of the theory. . . . "[16] The letter of 30 October gave an estimate of the expected rate of signal transmission over the cable and a calculation of income. In a brief report to the British Association in 1855, Thomson gave additional theory applicable to the cable and concluded: "Immense economy may be practised in attending to these indications of theory in all submarine cables constructed in [the] future for short distances; and the non-failure of great undertakings can alone be *ensured* by using them in a preliminary estimate."[17]

When Whitehouse, the electrical technician on the Atlantic cable enterprise, disputed Thomson's theory, Thomson, in an uncharacteristic move, entered into a public technological controversy aired in the *Athenaeum*. He wrote that Whitehouse's argument "not only profess[ed] to overturn [his] theoretical conclusions, but it gave what might at first appear to be sufficient experimental evidence of the validity of an ordinary submarine cable for telegraphic communication between this country and America, *in opposition to* [his] [Thomson's] warning. . . ."[18] (italics added).

By 1856, Thomson was engaged directly in technological questions. Although he was drawn in at first by the dispute over his theory, he step by step moved into strictly practical matters outside of the confines of scientific research. "In the mean time," he wrote, "as the project of an Atlantic Telegraph is at this moment exciting much interest, I shall explain shortly a telegraphic system to which, in the course of this investigation, I have been led. . . . "[19]

Public interest served as a vortex that drew Thomson in. Previously, he had managed to remain outside of the center of public concern, but the Atlantic cable drew him into the center of nineteenth-century technology. From 1854 to 1858, in spite of himself, William Thomson gradually became a central public figure and lost the anonymity of the theoretical scientist.[20]

Stokes's letter of 16 October 1854 (where he asked about the causes of the signal delay, which Faraday had investigated) caused Thomson to carry his mathematical analysis further. Instead of limiting himself to the calculation of one effect, capacity, Thomson now analyzed the cable as an electric-current carrying system and compared all the properties of the cable to Fourier's case of heat passing through a solid. The analogous elements in cables were resistance and capacity. In order to use Fourier's analysis, Thomson had to impose ideal conditions on the operating of the cable, for example, by assuming that the signal was applied in an infinitely short period of time and that the insulation of the gutta-percha was perfect. These ideal conditions were no more attainable in a telegraph system than they were in the passage of heat through a solid, as in the case where Fourier's analysis required the infusion of heat to occur in an infinitely short period of time and required that the solid be perfectly insulated from the surrounding air. However, as in the case of the Fourier analysis, where the abstract mathematical cause was related to actual conditions by making approximations after the mathematical solution was arrived at, in the case of the telegraph, Thomson assumed that a brief closing of the telegraph key was near enough to the mathematically ideal. Even though gutta-percha was not an ideal insulator, it was an excellent one, and under actual conditions the equations were sufficiently close.[21]

Why weren't Thomson's equations closer to the actual case the way Faraday's experiments were? His equations had to make assumptions such as considering a pulse of electricity as being applied in an infinitely short period of time. It was only by abstracting to these ideal conditions that the analogy between heat and electricity applied. Whatever uncertainty existed in Thomson's mathematical study of the cable was due to his assumption that the actual conditions were close approximations to the mathematically ideal. In Faraday's case, the question was whether his experimental conditions could be extrapolated to apply to the real situation. As it turned out, Thomson was more correct in interpreting Faraday's results than Faraday was.

On 28 October 1854, Thomson wrote to Stokes that all practical questions were answerable by the proper use of his equation. One characteristic of the Leyden-jar effect, or capacity of cables, was that the retardation of signals was proportional to the square of the cable length—a cable 1,000 miles long retarded the signal four times more than a 500-mile-long cable.[22] In an unpublished part of Thomson's letter to Stokes, he wrote that he had no more time for the subject and was going on to other research.[23] He was satisfied to let his analysis rest at this point in an idealized treatment of retardation. Limiting his interests to general theory was in keeping with his previous work on steam engines. He provided the theoretical base but not the blueprints for technology, and in this, Thomson went beyond what most other scientists were willing to do in the way of practical work.

Although it made specific reference to the Atlantic cable, the letter of 30 October gave an analysis that was applicable to any submarine cable regardless of length or dimensions. In this letter, Thomson derived a dynamical mathematical metaphor for cables. The equation treated cables as an electrical system and gave the relationship between the physical dimensions and the electrical characteristics. Given these measurements, the current at the receiving end of the cable could be calculated. Time, as in Fourier's analysis of the motion of heat, was a factor in Thomson's equations. The instant-by-instant magnitude of the received signal was calculable.

The tests conducted by Faraday determined the time it took a signal to pass through a cable, but what he measured was the time it took a detectable signal to pass. By simply opening and closing the electrical circuit, Faraday did not duplicate the pulsating conditions of an operating telegraph system, where not just one pulse of electricity was sent through the line, but dots and dashes followed each other in rapid succession.

Experiments treated cables as simple conduits with a pulse of current passing through them, and the object of the experiments was to determine the time for a signal to pass from one end to another. The metaphor of the mathematical equation depicted a cable in such a way that it showed the quantitative relationship between all factors involved in the transmission of a signal with the signal itself being represented by a quantity. The equation was a metaphor in that it described no cable exactly.

Thomson's picture of the signal arriving at the receiver showed a current whose peak was "worn down." Retardation, as he imagined it through his dynamical metaphor, resulted in a current at the remote end of the cable beginning at zero and gradually increasing in strength to a maximum value. The question raised by his equation was not when "a signal" was received but when "a sensible signal" was received. Thomson plotted curves of the received signals showing their gradual increase. For two cables with the same electrical properties and physical dimensions, the time for the signal to reach a stated fraction of the maximum current was proportional to the difference in length squared. In other words, retardation increased as the square of the distance.

In his letter of 30 October to Stokes, which was later published, Thomson claimed: "We may be *sure* beforehand that the American telegraph will succeed, with a battery sufficient enough to give a sensible current at the remote end . . . but the time required for each deflection will be sixteen times as long as [it] would be with a wire a quarter of the length, such, for instance, as in the French submarine telegraph to Sardinia and Africa." Thomson suggested testing the period of retardation in the Sardinia-Africa cable. Making the Atlantic cable equally efficient was a matter of increasing the proportions of the cable to compensate for the additional length. "It will be an economical problem, easily solved," Thomson wrote, given the prices of copper, gutta-percha, and the iron for covering the cable, as well as the message-rate needed—the minimum initial cost of the Atlantic cable was computable.[24] Thomson's analysis was the first time in which the operation of the telegraph was subjected to such careful theoretical scrutiny. The accuracy of Thomson's conclusions depended on the validity of his use of the analogy with the motion of heat.

One of the principal inferences that Thomson had drawn in his letters to Stokes, that retardation of the telegraph signal increased as the square of the cable length, was challenged at the meeting of the British Association during the summer of 1856 by E. O. W. Whitehouse, one of the original projectors of the Atlantic cable and its chief electrician. In his paper titled "The Law of Squares—is it applicable or not to the Transmission of Signals in Submarine Circuits?", Whitehouse stated:

> In all honesty, I am bound to answer, that I believe nature knows no such application of that law; and I can only regard it as a fiction of the schools, a forced and violent adaptation of a principle in Physics, good and true under other circumstances, but misapplied here.

Whitehouse scoffed at the practical implications of Thomson's theory. If the theory were followed in manufacturing the Atlantic cable, the cable would be too ponderous for any ship to carry. And the idea of perfect insulation—one of Thomson's hypothetical stipulations—was a practical impossibility, so that following all his recommendations meant that plans for the cable "would be abandoned as being practically and commercially impossible." Part of Whitehouse's impatience with the precautions urged by Thomson was that plans were well under way for the cable project.[25]

Whitehouse's comment that theory was "good and true under other circumstances, but misapplied here" was the practical man's expresssion of disdain for theory and implied that the theory had exceptions. From the scientist's point of view, that attitude undermined the very basis for theory. It was as if someone had claimed that Newton's universal law applied to all masses except the earth. To support his claim, Whitehouse supplied data from his own experiments that indicated exceptions to the law of squares. Thomson analyzed the experiments and interpreted the data. Where the data were not in agreement with the law of squares, he found the reason for Whitehouse's error.[26]

The controversy between the scientist and the practical man was on two different planes and was impossible of resolution. Thomson wanted the conditions for the experiments to be exactly controlled according to the principles to which Fourier applied heat. For example, Thomson pointed out that Whitehouse applied his battery to the telegraph line for one second, but Thomson's theory was based on an infinitely short application of electricity. Whitehouse had "disproved" the square law for very long cables by applying the battery for too long a period of time and not only did this confute the theory, but his method was not like ordinary telegraph transmissions where signals rapidly succeeded one another in the course of sending messages. In other words, Whitehouse's experimental conditions were artificial and unlike the conditions to be faced in the actual cable.

There were a number of points that Thomson raised dealing with the number and size of conductors as well as the receiver used by Whitehouse which made him question the feasibility of Whitehouse's plans for the Atlantic cable. In spite of reservations about Whitehouse's tests, Thomson said he was glad to see that Whitehouse had succeeded in convincing "practical men" that a cable of ordinary dimensions would work on the Atlantic project.[27] Thomson advocated a larger conductor and a larger cross section of insulation to compensate for the length of the cable.

In the unbalanced argument between theory and practice, Thomson did not press his argument because even though he thought Whitehouse was wrong about the theory, he thought Whitehouse might succeed in making the cable work. Whitehouse, on his part, was satisfied that he had won the argument. "It was upon this practical basis," he declared, "that I joined issue with Prof. Thomson, testing the application of his theory, as stated by himself, to the hard facts of practical telegraphic operations."[28] It was impossible to resolve this controversy as in the case of Faraday's dispute with the mathematicians when they differed over the explanation of the phenomena, that is, over the theory. Whitehouse, a practical man, had no theory; he only interpreted data.

He failed in his attempt to make the Atlantic cable work. It was William Thomson who designed a system that worked successfully. He was able to take the idealized mathematical theory, the mathematical metaphor, and correctly interpret it physically. It was not a case of applied science, as Thomson recognized, since his theory described a situation that was impossible to reproduce physically, for example, such things as an infinitely short pulse and perfect insulation. But if the theory had been less idealized, it would not have been applicable to all situations. Only the mathematical metaphor was general enough to cover the unforeseen. Thomson interpreted correctly the theory for practical application in much the same way he resolved the conflict between Faraday and the mathematicians, and Carnot and Joule, except that in the case of the Atlantic cable he was not performing as a scientist who resolved a dispute between two points of view—that is, he was performing the function of an engineer.

Thomson, in spite of himself, was drawn into the discussion of submarine cables, but he became an active force in building the worldwide network of cables as an engineer-consultant and as a holder of basic patents. Accepting his own view that technology and science are separate though related, his was not an easy transition from scientist to engineer, but a change. Consequently, his work was a dualistic combination of the two endeavors. In engineering, there were more direct rewards, a major one being money, and in this facet of his work, Thomson revealed an ambitious nature that resembled his father's.

Thomson's life up to this time, about the mid-50s, consisted of a devotion to scientific research that separated him from the mundane. His dedication to science is not understandable in ordinary terms of motivation, such as ambition, and this dedication remains one of the enigmas of Thomson's personality. Why this intense effort to develop theories and penetrate the core of ideas about such abstract questions as the nature of heat, of magnetism, or of electricity?

The ideas which he used to develop a scheme for operating the Atlantic cable originated before he had heard of the project or anyone had comtemplated such an undertaking. Tracing over the course of his thinking reveals how little his scientific research was done with the anticipation of any technological application. How devoïd of ambition, in a practical sense, were his efforts expended on his research.

Thomson's mathematical diary of 6 September 1850 recorded receipt of data taken by Professor Piazzi Smyth of underground temperature readings over a period of years. Thomson planned to use the data for deriving general equations for the passage of heat through solids with the earth taken as an example. Seeking a general expression for the motion of heat when a solid was infused with a quantity of heat in a pulselike manner, Thomson recalled notes he had taken in 1840 as he read Fourier's work for the first time. Starting with the core idea of Fourier's mathematical theory, Thomson's thinking began to cover more and more phenomena. The pages of the diary gave eight variations of the manner in which heat might be applied to a solid body including the example of an instantaneous shot.[29] When they were published, these equations were illustrated by diffusion diagrams that described the motion of heat, the diffusion of substances in solutions, and the movement of electricity along a submarine cable. These diagrams were the prototype for the curves Thomson drew for the retardation of current in submarine cables.[30]

When Thomson submitted his letters to Stokes to the Royal Society for publication, he asserted his intention of sending a paper about the telegraph with a more complete mathematical development of the problem. "Compendium of the Fourier Mathematics for the Conduction of Heat in Solids, and the Mathematically Allied Physical Subjects of Diffusion of Fluids and Transmission of Electric Signals through Submarine Cables" was not published until 1880, although, according to Thomson, the most important part was finished during September and October 1850.[31]

A similar course of his thinking can be found in his mathematical determination of the capacity for a submarine cable. Another of Thomson's papers,[32] an earlier theoretical explanation of Faraday's experiments on cables, was sent as an addendum to the reprint of Thomson's articles in the *Philosophical Magazine*. The following articles, "On the Uniform Motion of Heat in Homogeneous Solid Bodies, and its connection with the Mathematical Theory of Electricity" and "On the Mathematical Theory of Electricity in Equilibrium," contained principles, Thomson said, that enabled calculations to be made of various arrangements of wires covered with insulation, including the submarine cable used in Faraday's experiments.

The writing of this mathematical theory by Thomson, assiduously pursued in his diary over the years in spite of blind alleys and false starts, represents a mental effort of an extraordinary kind lacking the spur of ambition, that is, his reward of a professional chair already came at the age of twenty-two. Thomson could devote his time to his intellectual ambition because it was not necessary for him, as it was for so many of his contemporaries, to spend time seeking a suitable position. If he had not been awarded the chair, he would not have acted differently. However, having the chair at such an early age made him intolerant of those who appeared to grab at every opportunity. Although his work on the Atlantic cable did bring rewards, they were not consciously sought. Thomson, no doubt, was one of the aloof pillars of the scientific community that T. H. Huxley referred to when he wrote:

> In the autumn of 1851, my friend [John Tyndall] and I went to the meeting of the British Association . . . as scientific "items" not, indeed, wholly unknown to the "pillars" of that scientific congregation; and perhaps already regarded as young men whose disposition to keep their proper places could not, under all circumstances, be relied upon. Being young [Huxley was twenty-six, the same age as Thomson, and Tyndall was thirty-one], with any amount of energy, no particular prospects, and no disposition to set about the ordinary methods of acquiring them, we could conduct ourselves with perfect freedom. . . . [33]

Unlike Thomson, they were self-made men who saw themselves as outsiders. Their ambition was blatant, a kind of ambition that Thomson thought unseemly in a scientist.

When Tyndall published his paper on glaciers, he angered many of the "pillars" of the scientific community. Thomson was especially angry because Tyndall's theory of the nature of glacier motion and the structure of glaciers cast doubt on the work of his friend J. D. Forbes, and of his brother James. Thomson wrote to Nichol: "In fact I believe Tyndall means to be perfectly honest, and I doubt whether he is ever fully conscious of being artful . . . but as to motives I do not wish to judge."[34]

It was during his attempts to complete the Atlantic cable that changes occurred in Thomson's life, changes that brought about different attitudes and motivations. Yet his second career as engineer was related to his work as a scientist in more ways than the connection between the abstract mathematical theory and his patents. His motives for doing scientific research are not explained in terms of worldly gain. Once embarked on his engineering career, Thomson was quick to see the use of his ideas in improving technology and sought the rewards of money and reputation that went with those endeavors.

In December 1856, Thomson was elected a director from Glasgow for the Atlantic Telegraph Company. Field informed the board that he wanted the cable in operation by the end of the summer of 1857. Bright was appointed chief engineer and Whitehouse was named chief electrician. Before Bright's appointment was settled, the specification for the cable was set and the manufacturing contract was let. Whitehouse had approved the design, and his opinion was supported by Faraday and Morse.[35] The three agreed that the smaller the conductor, the better the transmission, Thomson's square law to the contrary notwithstanding.

When the cable squadron left Valentia Bay in August 1857, William Thomson was on board the *Agamemnon,* a converted British man-of-war that carried half of the cable. The other half was carried in the *Niagara,* the American Navy's contribution to the expedition. Whitehouse, chief of the company's electrical department, stayed behind at Valentia, Ireland, because of ill health. The decision to proceed in 1857 was contrary to Bright's advice. He urged a year's delay for conducting exercises in paying out the cable and maneuvering the ships as well as for further electrical experiments.

Two weeks before the expedition was to start, Bright was told that his decision to begin the operation in the mid-Atlantic had been countermanded by the board of directors. Whitehouse argued that it was essential for the cable-laying squadron to be in constant communication with the shore. The first splice would be between the shore end, which had the receiving instruments, and the cable ships offshore at Valentia. They were to pay out cable to the middle of the ocean where the second cable ship would splice its end and continue the trip to Newfoundland. Bright argued in vain that this meant doubling the time for work and that the splice then had to be made in the mid-Atlantic while one ship hung on to 1,000 miles of cable, making it impossible to wait for favorable weather.

Bright sent a long letter to the directors giving the mechanical and nautical considerations against the electrical ones. The electrical problem, he complained,

> must have existed from the first, and it is peculiarly embarrassing to me that the discussion of its importance—as compared with the other conditions necessary to success—should have been deferred until a fortnight previous to our departure, when we are under so great a pressure for time, if the undertaking is to be carried out this year.[36]

The only recorded suggestion that Thomson made was to Brett for using a hydraulic brake patented by Thomson and W. J. M.

Rankine. The expense of the patent, Brett said, was money thrown away, and with cool courtesy he assured Thomson that he would be happy to receive other suggestions from him, which would be received with consideration.[37] Thomson and Rankine allowed their patent to lapse.

After the cable squadron was 380 miles out of Valentia, the calamity that Bright warned against occurred. Too much tension had been applied to the restraining brake, and the cable parted in 2,000 fathoms of water. Bright was undaunted and reported to the directors that he was more than ever confident of success in the undertaking.[38]

Another letter dated 21 August 1857, addressed to the directors of the Atlantic Telegraph Company, appeared in the *Times.* It said: "For a cable of great length, necessarily limited to a single conductor, we cannot suggest, on electrical or any other grounds, any improvement upon the form adopted by the company." The letter was signed by Whitehouse, Morse, and Thomson.[39] Thomson still believed that the technicians might succeed although they worked on assumptions in opposition to his theory.

The general agreement was that the 1857 failure was due to the brake. A group of experienced engineers set about improving the brake design for the company between the summer of 1857 and the next planned attempt for the summer of 1858. Thomson contributed one piece of theoretical information. In a letter dated 16 October 1857 to the *Engineer,* a new journal, he gave the differential equations which described the curve of the cable as it went over the stern of the ship.[40]

Beginning with his publicly aired dispute with Whitehouse, Thomson made a number of suggestions for improving the operation of submarine cables by, as he said, working out various practical applications of his theoretical formulas. In December 1856, he presented a plan of operation for a telegraph system, including the method of sending a signal, receiving a signal, and a code for increasing the message capacity of the cable. The system, he confidently believed, was capable of sending a distinct letter every 3½ seconds, amounting to 17 letters per minute, or 200 messages of 20 words in a 24-hour period. At 30 shillings a message, that would be a good return for an investment of £1,000,000.[41] The instrument he proposed for receiving the signal was Helmholtz's galvanometer,

> with or without modification. The time of vibration of the suspended magnet, and the efficiency of the copper damper, will be so arranged, that during the electric pulse the suspended magnet will turn from its position of equilibrium into a position of maximum deflection, and will

fall back to rest in its position of equilibrium. The possibility of fulfilling
these conditions is obvious from the form of the curve I have found to
represent the electric pulse. The observer will watch through a telescope
the image of a scale reflected from the polished side of the magnet, or
from a small mirror carried by the magnet, and he will note the letter or
number which each maximum deflection brings into the middle of his
field of view.[42]

A year later, having "modified and simplified" the above plan,
Thomson developed his mirror galvanometer, which he patented in
February 1858.[43] The indicating medium was a light reflected from
a mirror that was attached to the movable element. The reflected
light shone on a surface a short distance from the galvanometer.
One of the features of the mirror galvanometer was the extreme
lightness of the movable portion which weighed less than half a
grain and was suspended by a single silk thread. It was designed to
give an indication at the smallest current and to respond to the
leading edge of the retarded signal; other telegraph receivers operated on much larger currents and reacted to the maximum strength
of the received wave—if at all. The sensitivity of the mirror galvanometer resulted from the extreme lightness of the movable element
combined with the amplifying effect of the light reflected off the
mirror so that the slightest movement of the galvanometer was mag-
nified in the movement of the light. The delicacy of the moving
parts of the galvanometer contrasted with the sturdiness of the hous-
ing to withstand rough treatment at sea. This combination of airi-
ness with sturdiness was often repeated in Thomson's later patented
instruments.

Thomson stopped sending letters to the *Athenaeum* and gave his
more specific ideas in papers to the Royal Society. Although he was a
member of the board of directors of the Atlantic Telegraph Company, he did not insist that the company engineers accept his ideas.
His attitude was that his calculations and suggestions might be useful
to those working on the project "if their experiments leave anything
undecided as to the best plan of cable for their purpose, and if they
have any confidence in scientific deductions from established principles."[44] Not one of these suggestions was used by Whitehouse, and
since the 1857 trial had not finished laying the cable, Thomson
found no cause to complain that he had been ignored because all the
electrical tests indicated that communication through the cable was
not a problem.

Purely by chance, Thomson decided to use some sections of the
1857 cable in experiments he was conducting on resistance measure-

ment. He was surprised to find a large variation of resistance in the samples. The cable's resistance, Thomson's theory indicated, increased the retardation of signals. The experiments showed that the higher resistance was due to a slight decrease in the purity of the copper used for the conductor. Thomson proved to the directors that by using a purer quality of copper, a cable with 50% less copper would have the same message capacity as a larger cable using less pure copper. He convinced the directors to specify high purity copper in their order for cable to replace the lost section. The manufacturer agreed to meet the higher specifications only after a premium of £2 per mile was offered. Thomson won his case "after much perseverance."[45] This was the first time he exerted an influence on the planning of the project.

An official statement from the Atlantic Telegraph Company published in the *Times* of 3 April 1858 announced the directors' intentions of consulting electricians and telegraphists in order to get their ideas for increasing the speed of transmission and improving the reception of signals through the new cable. Professor Thomson, among others, had been asked to consult with Mr. Whitehouse.[46]

Nevertheless, not much concern was felt for the ability of the electrical instruments to perform their part. In April 1858, when Thomson asked the directors for £2,000 to develop his mirror galvanometer, he was told: "The Directors, having regard to the reports and observations of Mr. Whitehouse, and particularly to the financial state of the Co., are of opinion that it would not be expedient to advance so large a sum under present circumstances. . . . " The directors did promise Thomson £500 for instruments.[47]

At the request of the directors, Thomson went on the 1858 cable-laying expedition and received no salary. Some concessions were made by the entrepreneurs. Bright took his experimental cruise on the Bay of Biscay before the actual operation began. As a condition of accompanying the expedition, Thomson insisted that he be allowed to keep his mirror galvanometer in the cable circuit during the laying operation, although Whitehouse's electrical equipment served as the primary instruments. The directors requested that Thomson take a more active part in the work. Did that mean that he was allowed more freedom? In 1859, an investigation was conducted in which Thomson testified that he "had comparatively few opportunities of investigating the state of the cable; indeed, I may say, I had no satisfactory investigation of the state of the cable, because it was actually on board ship before I was able to be on the spot, and joined up in such a manner that the testing of the different parts was not practicable." He complained that he had only fourteen days to conduct tests before the ships cleared the harbor.[48]

The ocean revealed itself as an unpredictable factor in cable laying. In June 1858, one of the worst storms known in the North Atlantic hit the H. M. S. *Agamemnon,* with Thomson aboard. The ship was an easy victim. She had 400 tons of cable in her hold, and on her decks were coils of cable that totaled approximately 500 tons. She also carried 160 tons of coal. After weathering the week-long gale without capsizing or sinking, the ship faced the climax of the storm:

> At 10 o'clock the Agamemnon was rolling and labouring fearfully, with the sky getting darker and both wind and sea increasing every minute. At about half-past 10 o'clock three or four gigantic waves were seen approaching the ship, coming heavily and slowly on through the mist nearer and nearer, rolling on like hills of green water, with a crown of foam that seemed to double their height. The Agamemnon rose heavily to the first, and then went down quickly into the deep trough of the sea, falling over as she did so, so as almost to capsize completely on the port side. There was a fearful crashing as she lay over this way, for everything broke adrift, whether secured or not, and the uproar and confusion were terrific for a minute; then back she came again on the starboard beam in the same manner, only quicker, and still deeper than before. Again there was the same noise and crashing, and the officers in the ward-room, who knew the danger of the ship, struggled to their feet and opened the door leading to the main deck. Here, for an instant, the scene almost defies description. Amid loud shouts and efforts to save themselves, a confused mass of sailors, boys, and marines, with deck-buckets, ropes, ladders, and everything that could get loose, and which had fallen back again to the port side, were being hurled again in a mass across the ship to starboard.[49]

Then the coals on deck broke loose. Several men were hurt but miraculously no one was killed or washed overboard. Thomson was unhurt, and even though the electrical cabin was flooded the instruments were undamaged, having been removed before the storm hit. The cable broke three times. Once, when Thomson obtained results that showed poor insulation somewhere in the 1,500 miles of cable, he begged the engineer to give him a chance to test a shorter length. It was necessary to slow the speed of the ship and clamp the cable, but the cable parted. Thomson had no doubt that the cable had failed electrically before it decided the issue and parted under the stern of the *Agamemnon.* The ships returned to port. Pessimism dominated the directors' meeting that convened in London. One group was in favor of giving up the project altogether and recovering as much as possible by selling the remaining cable. Three directors, Curtis M. Lampson, Cyrus W. Field, and William Thomson, won an argument in favor of another try.

The 1858 expedition was the most psychologically and physically trying of the cable-laying ventures. Thomson must have wondered what brought him to this pass. When he agreed to go on the *Agamemnon* in 1857, it was at the last minute and only because Whitehouse had backed out just before they sailed. Those on board the *Agamemnon* could not see land for thirty-three days in 1858, and Thomson was away from his wife for three months. It was a separation that was regretted by both.

The whole arrangement was highly unusual and promised no material benefit to Thomson. The men who accompanied the expedition, Bright and Field, were due to receive very large rewards, and their involvement was direct as engineer and entrepreneur. Thomson was a scientific adviser with no specific responsibilities and no authority. He followed Whitehouse's orders. But for him, the submarine telegraph was a personal challenge; it was a test of his theory. He threw himself totally into the affair and an observer might have believed he had much at stake. At one point in the laying operation in 1858, the signal failed and Thomson was sent for:

> He came in a fearful state of excitement. The very thought of disaster seemed to overpower him. His hand shook so much that he could scarcely adjust his eyeglass. The veins on his forehead were swollen. His face was deadly pale. . . . Dr. Thomson, in a perfect fever of nervous excitement, shaking like an aspen leaf, yet in mind clear and collected, testing and waiting, with [a] half-despairing look for the result. . . . [50]

The Atlantic cable marked the emergence of Thomson as a man who could think like an engineer. For a time, the expeditions seem to have liberated the dedicated scientist into a bold man of action. He took a turn at the laying-out brake when necessary and fell in with the companionship aboard. During games of whist, the other players were amused to see him act the role of an absentminded scientist often asking "Who played what?"[51]

At last, on 5 August 1858, a cable was completed from Newfoundland to Ireland, and the first electrical communication was carried across the Atlantic. Great celebrations took place on both sides of the ocean. All of New York City was illuminated. There was some delay while Whitehouse adjusted his instruments at Valentia. Soon the newspapers printed the text of the greetings exchanged by Queen Victoria and President Buchanan.

But there was trouble. Those connected with the laying of the cable knew from the electrical tests that there were several spots where the insulation was weak. Under ideal conditions, Whitehouse would have had trouble getting his less sensitive instruments to work. In his vain attempts to demonstrate that his apparatus was satisfactory, he applied very high voltages to the cable—2,000 volts at one time. He even raised the cable from the ocean floor to repair it. This last action was against the explicit orders of the directors. Those messages that did pass through the cable were received on Thomson's galvanometer.

Numerous investigations and a considerable amount of recrimination followed. Whitehouse was fired by the directors over the objections of Thomson. The vice-chairman, Lampson, wrote to Thomson: "I must not hide from you that the course you took in relation to our recent difficulties with Mr. Whitehouse added greatly to our troubles at a most critical period of the Companies affairs. . . ."[52]

In the aftermath of the affair of Whitehouse versus the Atlantic Telegraph Company directors, Thomson admitted that he knew that Whitehouse's conclusions did not follow from his experiments, "but I assented too readily to his own over sanguine expectations of what his instruments could do," he wrote to Nichol. "I put some fundamental questions to him which he answered so undoubtingly, that I thought he might possibly work with advantage by his instruments. . . ." Thomson did not stand against Whitehouse on the practical issues "because I [Thomson] was much better pleased to let the practical telegraphic work be carried out by others."[53]

No further attempt was made to lay an Atlantic cable until 1865. Meanwhile, several investigations were undertaken. One, conducted by a joint committee of the Lords of the Committee of Privy Council for Trade and the Atlantic Telegraph Company, after ten months of hearings, filed a report of over 500 pages, detailed and technical, that concluded: "We cannot but observe that practical men ought to have known that the cable was defective, and to have been aware of the locality of the defects, before it was laid."[54]

The committee reported that at the time, 1861, a worldwide total of 11,364 miles of cable had been laid. However, a little over 3,000 miles were actually working; nevertheless, "The failures of all these submarine lines are attributable to defined causes, which might have been guarded against. . . . We believe that there are no difficulties to be encountered in laying submarine cables, and maintaining them when laid, which skill and prudence cannot and will not overcome." Early successes in submarine telegraphy were a liability. "They became successful precedents to appeal to," the committee declared; "further investigation was thought unnecessary, and with no variation as regards the principles of construction, cable after cable was designed and laid down under circumstances and conditions having no resemblance to those originally encountered."[55]

In October 1863, the Atlantic Telegraph Company received the "Report of the Scientific Committee Appointed to Consider the Best Form of Cable for Submersion Between Europe and America." Thomson was part of that committee of five, and the report was a contrast with the hasty action of the company's directors in 1857. Specifications were made for conductivity of the copper, the size of the conductors, and for the insulation. These three factors were previously cited by Thomson as determining the speed of communication in the wire.

The directors of the company endorsed unanimously the recommendations of the committee. They did not flinch when it was found that the recommended cable required considerably more money, thus necessitating new financial arrangements. It was decided to form a new company, the Telegraph Construction and Maintenance Company, which would undertake the supervision of the manufacture and laying of the cable.

One other important change was made. Instead of depending on two ships with all the difficulties involved, the new expedition would have that seagoing wonder of the age, the *Great Eastern,* which took aboard over 2,000 miles of cable and more than 7,000 tons of coal. When the expedition sailed in July 1865, there were about 500 men aboard, and among them was Professor William Thomson. The electrical department was under the control of a man from the Telegraph Construction and Maintenance Company, and Thomson represented the interests of the parent company, the Atlantic Telegraph Company.

Electrical faults, due to pieces of the outside iron wire sheath which pierced the conducting core, plagued the work. Several times the *Great Eastern* had to stop and pick up a portion of the laid wire to make repairs. After 1,200 miles of the cable had been laid successfully, the command was given to start picking up the cable in order to check a suspected fault. While this process was going on, the cable parted and disappeared over the side in 2,500 fathoms of water. As the engineers and crew were close to bringing the lost cable back on board, they ran out of grappling equipment and had to return.

For the fourth time, an Atlantic cable-laying expedition returned to Britain to announce its failure. Another series of negotiations was started to raise capital for the fifth attempt. To raise the capital, it was decided that yet another company had to be formed, the Anglo-American Telegraph Company, capitalized at £600,000. The new company contracted to manufacture and lay down a cable in 1866.

Within a period of two weeks in July 1866, and with only one interruption in the work, the *Great Eastern* finally succeeded in putting down a well-functioning cable. No sooner did the ship touch Newfoundland than it turned around, grappled for and raised the 1865 cable, and then completed that line too.

All of these efforts and failures did not dampen public interest in the cable. The *Times* of July 1866 carried one of the best eulogies to technological progress ever written. The cable was, the paper said, "the most wonderful achievement of this victorious century." As a mechanical feat, the cable was a wonder, but the use of electrical science produced something much more mysterious. "When we see a mirror wavering and flashing in a little room in Ireland," the newspaper declared, "what imagination would ever have dreamt that the movement was caused by a slight and silent operation on the shores of Newfoundland? . . . The two worlds have formerly been so widely separated. . . . America cannot fail to live more in Europe and Europe in America. Nor is the effect of this close communication only to be anticipated in a keener sympathy and a closer relation; it must be seen in a quickened and more energetic life. . . . For the purposes of mutual intercourse the whole world is fast becoming one vast city."[56] On 10 November 1866, along with other projectors of the cable, Thomson was knighted by Queen Victoria. His contributions to scientific theory earned him the admiration of his colleagues long before he became Sir William Thomson. But he was little known outside of the scientific community until the cable brought him a popular success which would last as long as he lived.

Thomson was a successful engineer to the British people, and few knew how recently he had entered that profession. The formation of a partnership with two other engineers, Fleeming Jenkin and Cromwell Varley, with the purpose of exploiting their telegraph patents, was a manifestation of the new Thomson. With Jenkin, he also formed a firm of consulting engineers. They collaborated on several important submarine telegraph projects.

To some observers, William Thomson may have seemed a remarkably quick and mystifyingly successful scientist. To those ignorant of the unifying themes in mathematics and such concepts as the dynamical theory, which gave his work a basic coherence, his thinking was deemed hopelessly unconnected. He never seemed to follow through, although he took each topic as a piece of a larger generalization. His papers were brief answers to short questions rather than meandering threads to be carefully traced. He worked at odd moments for short periods and worked best when away from home and traveling. Joule was frustrated in his efforts to have Thomson concentrate for any period of time on their research. Thomson always had other interests.

The Atlantic cable commanded a long period of undivided attention. What had begun with a question by Stokes and a controversy with Whitehouse became an unshakable interest. In previous scientific controversy, after stating his view, Thomson was satisfied to await developments because "in the present state of science" nothing more could be said. In the case of the cable, the test of the theory was imminent. The promoters were determined to see the project through—if Thomson wanted his theory tested, he had to participate. Field's mission was to push the project through in the least amount of time. The system had to be workable, not necessarily the best possible.

Thomson was borne along by the momentum of the project and was not allowed to lay aside the question. He learned what it was to make engineering decisions; one of the requirements was to make them in time. From an aloof scientist willing to give advice, he became a planner and administrator of some of the most important technological projects to develop in the nineteenth century.

CHAPTER NINE

"The next generation will thank us."[1]

William Thomson and his wife spent the Christmas holiday of 1860 with friends in Largs. Margaret Thomson came down to Largs about a week before her husband, who came on 21 December. The holiday was cut short when William broke his leg on the ice the day after he arrived. He and his host had gone to a frozen pond to play a traditional Scottish game, curling. An accident occurred when a board Thomson was standing on hurled him onto the ice and injured his leg. Because of the indecision as to whether there was a break, the bone was not set until a week later. He had a bad fracture. The neck of his left femur sustained the whole momentum of the fall. Thomson endured a great deal of pain during the setting of the fracture, although large amounts of chloroform were given to him. His convalescence lasted until the fall of 1861.[2] He was still in a weakened condition the following winter. For a time, it appeared that Thomson would be an invalid like his wife, but the only disability of any permanence was a decided limp.

During the early part of his convalescence, he was forced to remain completely immobile and was not even allowed to raise his head, a condition harder to bear for a man such as Thomson than the concomitant pain. By the end of February, Thomson was eating and sleeping fairly well, but he had made little progress and was very thin. Despite his becoming "easily fatigued by any mental exertion,"[3] Thomson had an active correspondence with Fleeming Jenkin on various problems connected with engineering, especially cables. Four months after the accident, he wrote to Jenkin that his leg was still virtually useless.[4] When would he regain his strength? Would he ever have the use of his leg again? Would his active life turn into a sedentary one? It was thoughts like these, no doubt, that led Thomson to write to Jenkin:

"The Next Generation Will Thank Us."

What would you think of compounding a book on the Telegraph out of eligible portions of our patent and other inventions, and my telegraph article &c &c. The latter Sir C. Bright will let you have (a revised proof I lent him) if you desire it. Let me know and I shall write to him or you can ask him if you see him. A chapter on electric communication ought to be included. I had nearly written one for the Encyclopaedia article, but I was confined to too narrow limits to include it. I have written the substance of it to you. When I complained of limitation of space, the publishers, Black, said they would be glad to publish a book, of a sufficient comprehensive bulk, on the subject, but I feel . . . too slow and fatiguing (to myself I mean) a writer to tackle such a job. If we were to undertake a joint book, I could not undertake as my share more than what is specified above.[5]

Thomson wanted to write a book, and when Peter Guthrie Tait came to Largs to visit,[6] it must have occurred to Thomson that if his plan to do a book with Jenkin aborted, which is what happened, Tait would be a good coauthor.

In the midst of continuing problems with the Atlantic cable, William Thomson took on another work—a textbook. The concept for the book was Tait's, and he thought he would probably do most of the writing because he saw that Thomson felt "a repugnance to it which is not common." Evidently, Tait hoped for great advantage from the collaboration on this book because he wanted Thomson to cooperate next on a "great" work, a mathematical treatise about natural philosophy.[7]

Like Thomson, Tait's early education was in Scotland. (He studied at the Edinburgh Academy the same time as James Clerk Maxwell.) He went to Peterhouse in Cambridge, where William Hopkins was his tutor. But he bettered Thomson's record by becoming senior wrangler in 1852 and equaled it as first Smith's prizeman. Thomson came to know Tait in approximately 1860 when Tait became a candidate for the chair of natural philosophy at Edinburgh. James D. Forbes, a close friend of Thomson, resigned the chair because of ill health. The list of candidates, containing the names of some of the best scientists in Britain, was finally narrowed down to Maxwell and Tait. Both men showed unusual promise, but Maxwell led by virtue of publication of more significant work. If the position had been in an English university where little lecturing was done, the selection unquestionably would have been Maxwell. The appointment, however, went to Tait. The *Edinburgh Courant* explained that although Maxwell was

already acknowledged to be one of the remarkable men known to the scientific world . . . there is another power which is desirable in a professor of a University with a system like ours, and that is, the power of oral exposition proceeding upon the supposition of a previous imperfect knowledge, or even total ignorance, of the study on the part of pupils. We little doubt that it was the deficiency of this power in Professor Maxwell principally that made the curators prefer Mr. Tait.[8]

Without question, Tait was one of the best university lecturers on science of his time. His lectures "were always models of clear and logical arrangement. Every statement bore on the business in hand; the experimental illustrations always carefully prepared beforehand, were called for at the proper time, and were invariably successful."[9]

The need for a clear, well-organized textbook to accompany his lectures was what prompted Tait to start the project. He began negotiations with Macmillan for publication and about that time was pleasantly surprised by Thomson's offer to collaborate. Once the project was undertaken, the energetic Tait was swept along with the work. His first letters to Thomson in December 1861 were eager and hopeful. An eight-page letter discussed experiments on radiant heat, a mathematical demonstration of Green's theorem, as well as some ideas about their treatise, and ended with: "I am myself a good example of the want of such a book as we contemplate, having got all my information bit by bit from scattered sources, which often contained more error than truth." The treatise was meant not only for the classes of Glasgow and Edinburgh, Tait declared, "the next generation will thank us."[10]

Determined to avoid giving details of a subject that was rightly a part of classroom instruction, Tait wanted as much explanation as possible, "but *not* elaborate detail." In order to avoid the mistake of devoting four-fifths of the space to details, as a standard text of the time had done, he thought they might limit their treatise to three moderate-sized volumes, giving a far more complete course in mathematical and experimental physics than was available in French or German. There were none to be had in English.[11]

On Christmas Day 1861, Tait sent more details of his proposal. There were to be at least two volumes of the first work, the experimental one. Each was to write alternate chapters and forward his draft to the other, who would annotate freely. Tait emphasized the need for extensive comments in order to blend their styles artistically. He also saw this was the only way to iron out differences between them on theory so they would have a single point of view. A list of the subjects that occurred to Tait as he wrote showed "[they had] no light undertaking before [them]." But he was certain that by working three or four hours a day, nothing would hinder the completion—"or at all events [have it] in a state in which I can manage it alone"—of at least the first volume by May 1862.[12]

Thomson had no taste for long projects. At one point, he suggested that they hire a young man to take notes in their natural philosophy classes in order to hurry things along. Thomson wanted to signal when the recorder was to begin and when the recorder was to stop. "It often happens that about ¼ of one of my lectures consists of statements which would be useful to have in writing: the rest being chaff."[13] Nothing came of the idea, fortunately, because a recorder would have difficulty following one of Thomson's lectures.

Even when he gave a formal public lecture, such as the Rede Lecture at Cambridge in 1866 titled "The Dissipation of Energy," the audience became hopelessly lost in the pursuit of his thinking:

> The thread of the discourse was incessantly interrupted by digressions. As soon as he had got a little under weigh, something apparently unconnected with his subject would occur to his mind, and he would remark: "When you meet with a fallacy in vogue it is well not to leave it alone without having a rap at it" (or something to that effect), and thereupon proceeded to demolish the fallacy. That done, when both he and his audience had forgotten where he left off, he resumed the track, but not quite where he had left it, and by the time he and his audience had got together again, some new hare crossed his track, and we were all scattered in pursuit of it; and so it went on, the time running out, and the shortcuts to regain the track becoming more and more difficult to follow, until the end. The hunt pleasant enough to those who were nimble enough, but the game killed mice and rats—no hare.[14]

Neither man foresaw the main difficulty: How should the arguments for the new approach to science be presented, while avoiding the popular fallacy of assuming that the new theories were settled facts? Tait thought the *"only* difficulty" was the order and arrangement, but once they had *"very carefully"* worked it out, the rest would be "easy." He proposed organizing their book in the same way as his lectures, which was not very different from the accepted approach.

Tait was exasperated with Thomson's additions to the first outline draft and complained that if they went into that much depth, an additional volume would be required. What was needed "at once," Tait wrote, was a textbook for elementary teaching. He proposed that they write a "thoroughly trustworthy" popular book first, suited for general educational purposes, and then "astonish the world" with a hitherto unattempted book for a complete course in natural philosophy, both experimental and mathematical. But first, the popular book, and it had to be inexpensive. "We may mulct and bleed Oxford and Cambridge and Rugby &c, &c, to any extent, but how about our own classes?" In spite of these convictions as to the level of the book, Tait was willing to go along with whatever Thomson thought, as long as they accomplished something.[15]

Tait wanted the treatise to be experimental and mathematical, but it should maintain the traditional dichotomy. The two men were at home interpreting experimental data as well as in working physical questions mathematically. Thomson passed easily from one point of view to the other. In several instances of major importance, as in the case of reconciling Faraday with the Continental mathematicians and in that of Joule and Carnot, Thomson was able to combine the experimental with the mathematical. However, it was not a habit of thought. It was natural that the coauthors perpetuate the dichotomy between experimental and mathematical because they were taught that way.

When Thomson attended lectures on natural philosophy at the University of Glasgow in 1839, he was told that the study of science was divided into pure science and physical science. "In pure science the phenomena are deduced mathematically," Thomson wrote in his school notebook. "In mixed [physical plus pure] the results are deduced from experience, and applied by mathematics to more extended cases." The class studied mechanics, the most mathematical and therefore the purest of the sciences, and they touched on heat, electricity, and magnetism—all were taught by means of the instructor's demonstrations. They were told, "Some branches are almost wholly experimental, some, of which the ultimate laws are known, can be treated [as] almost wholly mathematical."[16] It is certain that Tait heard very much the same thing when he was a student at Edinburgh.

At Cambridge, Thomson and Tait studied under Hopkins and were required to copy and study Hopkins's unpublished manuscript. The approach was wholly mathematical, covering such topics as lunar and planetary theory and finishing with hydrostatics and hydrodynamics. In Thomson's copy, made in 1843, the manuscript has a section on the passage of fluids through tubes under the general heading "hydrostatics." Hopkins's purely mathematical treatment began with an assumption about the motion of a fluid particle that was "only . . . approximately true." No consideration was given to such dynamical considerations as cohesion, friction, or eddies.[17]

When the topic of fluids was broached in the treatise, the authors recognized the change in thinking. In his preliminary remarks for a section of the treatise concerning friction, Thomson speculated on the role of viscosity in the forming of a bubble film. It was very difficult to tell why bubbles appeared on the surface of pure water. The appearance of these bubbles was impossible according to the theory of viscosity. "In fact," Thomson mused, "bubbles are a great wonder every way; while notwithstanding, they give such very definite dynamical problems to solve as to their temporary equilibrium." Then he added, "Don't assume all I have been writing about froth [bubbles] to be correct. I have only been trying to see or think through difficulties."[18]

The separation of experimental from mathematical was restated in Thomson's introductory lecture at Glasgow, given first in 1846 and repeated at the beginning of each year with amendments and deletions. The subject of natural philosophy was dynamics, "the *science of force.*" Dynamics was not only basic to the study of natural philosophy but, because of its completeness as a science, it held first place. Thomson spoke of a few simple, axiomatic principles founded on common experience and expressed as general laws used for describing all dynamical action. The deductive processes of mathematical analysis applied these laws to particular cases. "Hence it is," his classes learned, "that dynamics is said to be a branch of mixed or applied mathematics." In the study of heat, electricity, and magnetism, not as far advanced as mechanics, observation and experiment were still the principal sources of knowledge. "Hence what is called the experimental or physical course includes these three subjects; while the more perfect sciences of mechanics and optics, being really mathematical subjects, form a distinct division of the studies prescribed by the University for a complete course of Natural Philosophy."[19]

Tait had a slightly different point of view. In his inaugural lecture given in 1860, he said that natural philosophy, "unlike Mathematics, is strictly a science of observation and experiment, or, as we may call the two combined, experience." Mathematics was not concerned with "any actually existing things, but simply with ideas. . . . There never was, and never will be, a straight line or a circle, still less a parabola or a cycloid." The object of experiment was to discover the laws of uncomplicated effects found in an intricate world of nature and, having been found, the laws were embodied in mathematical language. He continued, "The science has then reached a point at which it is needless to try experiment, the powers of mathematical analysis being sufficient to guide us to a result. . . . Few branches of Natural Philosophy are as yet, in this condition, or even nearly so," Tait told his students. Mechanics was one of those few branches. Because of this dual aspect of natural philosophy, Tait believed, "neither mere mathematics, nor mere experiment, will ever make a complete physicist."[20]

Tait, an accomplished user of experiments for classroom demonstrations, was a booster of Thomson's inventions of laboratory instruments. He found many uses in the classroom and laboratory for the mirror galvanometer, which he urged others to adopt. Tait often had a beneficial influence on the course of Thomson's inventions. One, a portable electrometer, was designed for taking readings of atmospheric electricity. When Thomson urged Tait to buy one for himself, Tait replied that he was waiting for Thomson to design a smaller electrometer, the size of an orange that would fit into his pocket. That prodded Thomson into developing an improved electrometer that did fit the pocket.[21]

Although they agreed fundamentally on the approach of the book, some issues had to be resolved in interpreting specific topics. Who won these arguments, and how did they determine which view would prevail? Much depended on the authors' personalities. Tait was more aggressive, but he tended to defer to Thomson's greater experience. Thomson had worked longer on the more fundamental questions. Tait thrived on controversy; it suited his peppery manner. He even wanted to settle some scores in their preface. "What do you say," he asked Thomson, "to a shot at the Critic of Faraday? Such as I have hinted (in an interlineation) without mentioning names, about that philosopher who is *so* great that other men's discoveries become his as soon as he repeats their experiments."[22] Tait meant John Tyndall, for whom he had an unremitting dislike. There were several bitter public arguments between the two.

Although Tait was usually the aggressor, when they attacked Tyndall in 1862, Thomson found himself out front. Tyndall gave a lecture at the Royal Institution titled "On Force." He credited J.R. Mayer, a German physician, with finding the mechanical equivalent of heat in 1842, a year before Joule. Both Tait and Thomson were angered by what they considered to be an erroneous historical interpretation of the development of the theory of heat, and the argument became public in the pages of the *Philosophical Magazine*. Thomson and Tait wrote an article titled "Energy" for the popular journal *Good Words* in which, as Tyndall said, they asked the readers "to decide between the rival claims of Joule and Mayer, and to form an opinion as to the scientific morality of myself."[23] Tait replied, "We seized the opportunity of distributing among [the journal's] 120,000 readers a corrective to the erroneous information which we saw was stealing upon them through the medium of *popular* journals."[24]

Tyndall's reply was addressed to Thomson and not Tait because "you are the older and more famous man, and it is your behaviour in this controversy, and not that of your colleague, which will interest the scientific world." Thomson's haughty reply was not convincing and did him no credit.[25]

When H. E. Roscoe's article titled "Thermo-Dynamics" appeared in the January 1864 issue of the *Edinburgh Review,* Thomson wrote to his brother James that it may have indicated that Tyndall's influence was "very decidedly on the decline, if not absolutely gone. There can be no doubt but that our late war with him has had good effect; possibly on him and certainly on the public."[26] Thomson was deluding himself. Tyndall's influence on the popular view of science was strong.

In the winter of 1861–1862, a parliamentary commission inquiring into the operation of the University of Glasgow proposed an ordinance that would stop payments of a graduation fee to the professor of natural philosophy, Thomson. The loss of income from this ancient right would mean the end of the natural philosophy teaching laboratory at Glasgow. As Thomson explained in his letters to the university commissioners, he used the income from graduation fees to pay an assistant, Donald Macfarlane, who not only helped with the classroom illustrations but supervised the research being done by students in the laboratory. In addition, Thomson was still recovering from his accident and could spend only a few hours in class, thus necessitating an assistant more than ever.[27]

An assistant was an ideal adjunct for the restless, dynamic Thomson. He had an eagerness for new ideas and an intellect that welcomed innovating concepts. As Joule and Jenkin learned, Thomson did not balk if these concepts upset his preconceptions. Thomson's light grip on ideas might have resulted in considerable loss, save for the patient and careful assistance he received from coworkers and laboratory assistants. Many of Thomson's ideas were suggestive, requiring supporting data before anyone would accept them, and without the support of the data of laboratory experiments, these ideas were in risk of being lost.

Fleeming Jenkin wrote that he hoped Thomson would "bring down [the university commissioners] without using powder and shot." Thomson barely used feathers. But he did threaten to stop the teaching of research and let the natural philosophy class revert to the method used by his predecessor. He even planned to petition the Queen in Council, but decided against it because, as he told Archibald Smith, he had little faith in the commissioners' promise of fair consideration.[28] Had Thomson's father lived, the fight with the commissioners would have been pushed more vigorously. Had Tait been in Thomson's place, he would have persevered in the battle.

The growth of the natural philosophy laboratory took place gradually after Thomson's appointment in 1846. He was allowed a college servant to help in the setting up of class illustrations, and the faculty granted him additional funds for purchasing new apparatus so that the laboratory equipment was in keeping with the advanced state of natural philosophy in 1846. That arrangement was all that was expected, but "I was anxious to take part myself," Thomson told the commissioners, "if possible in advancing science, and therefore after having once thoroughly learned the use of the instruments and the manipulations of the experiments required to illustrate my lectures, I commenced employing, at my own private expense, assistants to prepare these experiments, and devoting some of my own time to a higher kind of work."[29]

There was a gradual increase in the involvement of the students in original research. Thomson wrote to James in 1855, "I have been more fully occupied with my students this session than almost ever before, as I have a number (sometimes more than a dozen) working in the apparatus room, carrying out *original researches* for several hours every day. Two or three decided and very good results have been obtained."[30]

In 1860, when the Royal Society rejected an application from Thomson for a research grant on the grounds " 'that it seemed rather for carrying on the general experimental work of my laboratory, than for some specific object involving considerable outlay.'— This is a most erroneous impression," he wrote to Stokes. He explained that although the primary work of his laboratory was the preparation of the lecture illustrations used during the winter months, "I have besides instituted a system of experimental exercise for laboratory pupils in which I am induced to persevere; devoting a great deal of time to it, and a larger expenditure out of my private resources than I feel to be altogether consistent with other claims." He usually had about twenty such laboratory students during the winter months, "all volunteers and paying no laboratory fee," and of these students, several were usually found who were "efficient for original investigation." These were the young men employed in the original work for which Thomson requested Royal Society funds. He also kept his laboratory open during his six-month vacation "exclusively for the prosecution of original investigations." During those months, Macfarlane, who was paid out of private funds, that is, partly from Thomson's graduation fees, was occupied solely in assisting in Thomson's research and in carrying out his directions "when I am not on the spot." During the summer, Thomson usually found a competent student who worked as a second assistant.[31]

In this gradual and unofficial way, Thomson had founded a teaching laboratory, not only for instructing students in the methods of physics research, but also for introducing a few exceptional young men to original research. His was the first laboratory of its kind in Britain and, as he told the university commissioners, he foresaw that his laboratory would be a model for others to follow.

In spite of Thomson's efforts, the commissioners terminated the archaic system of fee paying. Instead they passed an ordinance giving Thomson a yearly sum obtained from state funds for operating the laboratory. Thomson's reputation as a scientist earned him the consideration that his demeanor did not demand. The arrangement was most satisfactory to Thomson. In November 1862, he datelined a letter to Stokes "Natural Philosophy Laboratory" and explained it was "a new institution of a most satisfactory character, which I have commenced with this session, owing to [the] University Commissioners' grant of assistant, and the College, proper rooms &c."[32]

Meanwhile, Tait pressed on with their book that was progressing as well as he wished. During January 1862, letters passed between the two men every two or three days. Both men were hard workers, but Tait concentrated more on the book. Although Tait intended to revise his chapter on the properties of matter, he was uncertain about Thomson's ideas because his notes on the manuscript were not clear. Tait sent Thomson his first draft of the preface and promised to fill Thomson's request for a draft of "abstract mechanics" soon.[33]

The complexity of the mathematical and the experimental in scientific theory and research confronted Thomson and Tait when they decided to include both of these viewpoints in their book. However, which point of view for each topic was still to be determined. Heat, electricity, and magnetism were no longer experimental subjects as they had been when Thomson was a student at Glasgow. Fourier and Carnot had devised a mathematical theory for heat. Green, Poisson, and Thomson had done the same for electricity and magnetism. Joule's and Faraday's experiments had changed the conception of the nature and mode of action of thermal and electric forces.

The universal and distinct nature of energy was to be the mainstay of their textbook. That energy, its conservation and convertibility, was a fundamental aspect of all phenomena was an idea promoted by experimentalists. Mathematicians previously treated each form of energy as a separate entity. Once Joule successfully promoted his idea of convertibility and equivalence, the mathematicians, primarily Thomson and Maxwell, developed more general mathematical theories which generalized the knowledge that was gained experimentally. In this case, the course of events in the evolution of the theories was from the experimental to the mathematical. Advanced researchers in heat, electricity, and magnetism needed to know only the mathematical theory. In order to understand these phenomena, students, in effect, needed the historical approach which showed the development of theory. Both the physical and experimental concepts of Joule and Faraday, as well as the later mathematical ideas were necessary for understanding the basic concepts.

The mathematical theory dealing with energy was the second significant development of physics, and by tacit agreement between the two authors, it also needed to be described in an introductory textbook. In the phenomena of heat, electricity, and magnetism, Fourier, Green, and Thomson selected mathematically describable factors to develop equations for describing and predicting the effects caused by these different forms of energy. The manner they used to do this became an integral part of the new physics. In this mathematical method, the ultimate cause of these forces was not sought. The theories of heat, electricity, and magnetism were able to predict accurately effects not yet discovered experimentally. Maxwell's equations predicted the whole realm of wireless transmission. But that proof was yet to come when Thomson and Tait were working on their treatise.

The experimentally established idea of the dynamical nature of phenomena plus the mathematical mode of representing energy were two inseparable concepts. The importance of the Thomson and Tait treatise was that they accepted the new dualistic nature of science. Even though they, too, spoke of the mathematical as being a more mature stage of theory, by their actions and the way they organized their textbook they helped to inculcate the next generation with the new concepts and methods developed by two generations of scientists from 1820 to 1860. The basis for the new approach to scientific theory and research began with Fourier and Green in the 1820s, included Faraday's work in the 1830s, Joule's research in the 1840s, Thomson's innovations in the 1850s, and was capped by Maxwell's work in the 1860s. A textbook that encompassed this era would be a sound basis for studying the new physics.

What needed to be changed in the treatise was often a matter of definition or of presentation. In commenting on Tait's draft of the chapter on abstract mathematical mechanics, Thomson asked if "Dynamics" might not be a better title. Tait saw advantages in the change because first of all, there was no such thing as statics—only dynamical equilibrium—and second, dynamics really meant the science of force or power. He said, "I am perfectly willing to drop mechanics entirely, and make Dynamics the general title. What would you propose as a substitute for the phrases *mechanical equivalent of heat* &c?"[34] The term "mechanical equivalent of heat" was well established through usage, and the authors decided not to change it.

When Thomson wrote the section on friction and cohesion, he remarked that "statical friction" was no longer an acceptable term. "To say force is not *dynamical* but statical is nonsense," he wrote to Tait, "except when dynamics is used in the corrupt sense from which we extricate it."[35]

"The book has been dragging its slow length along more slowly than you could conceive," Thomson wrote to Helmholtz. The burden of compiling such an inclusive textbook irked him, and he often complained about the demands of "the book." The chore was more to Tait's liking for he was closer to being a popular expositor of science than Thomson. Tait was willing to work for hours at a time with a pot of beer nearby and only his pipe for company. He gladly gave up dinner parties, which bored him, because he could rarely smoke or talk about science. Golf was one vice he did not give up, and he took an occasional day off from book writing to spend a day on the links. Evenings Tait gave entirely to the book except for marking examination papers of which he had 120 sets, with 12 questions each, every two weeks. How lucky Thomson was to have an assistant mark his papers.[36]

Of the two, Thomson was more interested in the fundamental and speculative questions. When it came to such ideas as the property of ultimate matter, Tait admitted: "You have evidently thought more deeply about matter than I have. . . . " But there were challenges by Tait as in the case of Thomson's theory about the ultimate compressibility of molecules.[37] It was Thomson who worried about such things as the cellular structure of matter. "By cellular I mean intensely heterogeneous as to density, with continuity through all the densest. [W]hat a contrast to the common idea of atoms, or even of molecules!" Thomson thought about but never wrote these speculations, although, he said, "I have hinted at it over so many years in my lectures, but of late, the compulsion to think for the book has made me feel it to be of more importance than it seemed to be before. I am persuaded that it has some positive truth in it."[38]

In an anecdotal aside, Thomson talked about the difference between naturalist and physicist. "Physician" was a better name, but "the doctors have monopolised it." The true distinction was to call men who described nature "naturalists," and those interested in dynamical questions, "natural philosophers." The description was applicable to books also. There were books of descriptive science and books of dynamical science, and of course their treatise fell in the category of dynamical science.[39]

Together, Thomson and Tait were incisive and not a little daring. When established physics in the form of Newtonian thought obstructed them, they undertook to place the *Principia* in a nineteenth-century context. Thomson believed that a fundamental difficulty with Newton, from the dynamical nineteenth-century point of view, was the question, What is force? One of Thomson's comments on their first draft was:

> Alas, alas. Behind all is the convention "what is force." It too is merely relative; but in our 1st Edition perhaps we had better prop up the tottering and doomed systems which make it absolute.[40]

Before they discussed Newton's *Axiomata,* they decided to define an axiom. On the draft Thomson had written that "an axiom is a proposition the truth of which must be admitted as soon as the terms in which it is expressed are clearly understood." Here Tait added in red ink: "by *those who have experience* in the subject *only.*"[41] The published version read: "Physical axioms are axiomatic to those only who have sufficient knowledge of the action of physical causes to enable them to see at once their necessary truth."[42]

The authors had a fundamental disagreement over Thomson's use of the term "interstellar air." Tait wrote back, *"That* is one of the great stumbling-blocks between us as joint authors. I can't rightly appreciate your idea of an unlimited atmosphere—I have seen *hints* of it in your papers, but *no reasoning.*" Tait suggested that they say simply "matter," and since it made no difference to Thomson, the term used in the book was "matter."[43]

Their method of work, sending drafts back and forth through the mail, tended to blunt their sharp remarks about each other's ideas. But the delays began to wear on Tait. Thomson's drafts were written in his green books, so-called because of their green covers, and were a continuation of his mathematical notebooks. He began to use them when he was flat on his back recovering from his accident.[44]

The negotiations with Macmillan continued. Should they publish a cheap edition of the book with smaller type and thinner paper? Macmillan opposed the idea because it would jeopardize the sale of the more expensive edition. In February 1862, the publishers were planning advanced advertising and even prepared the title page.[45]

The *Treatise on Natural Philosophy* was published in 1867, almost six years after Tait's first outline.[46] The writing tried Tait's patience. After one of Thomson's delays, Tait's letter began: *"Do* look alive. . . . "[47] Although the textbook was a major project for them, they were always occupied with other research and publishing projects. Thomson was improving his atmospheric electrometer and seeing to it that models were distributed to observers around the world, working on submarine telegraphs, and entering into a dispute with geologists on the age of the earth. In 1866, he gave the Rede Lecture at Cambridge. On top of these activities, he often went to the watering spas on the Continent during the summer months. But now these trips were for the benefit of his leg as well as his wife's health. Thomson's travels increased the difficulty of communication at a time when he and Tait did most of their work. Tait had his own work to do as well. In 1867, he finished his historical sketch on the dynamical theory of heat and a work on the new mathematics of quaternions. Tait's concern for the treatise during six years of anxious effort was unremitting.

During the early 1860s, Thomson was an exceedingly busy man. No one, including Tait, would have denied it, but Thomson's procrastinations had another cause. In May of 1864, Tait wrote: "Your recent comments on the attraction of an Ellipsoid are monstrous—as you had the M.SS. in your hand for a week, and I made all the alterations you indicated *then*. . . . " Doing things over was too much for Tait, and he foresaw no end to the delays. He demanded that Thomson "send me my M.SS. and then go and see Macmillan and account to him for deficits and extortions &c &c and promise him on *your* part . . . that Vol. I will be ready in the end of July."[48] For Thomson, the changes were the result of rethinking questions that were on his mind for a while. By putting his thoughts in writing, he revealed how incomplete some of his ideas were. He believed that all physical questions were related, and in bringing all those subjects together, he recognized that there were inconsistencies and disconnections in his thoughts. As Helmholtz had stated, "writing a book helps the author discover gaps, not so much in science but in his own thinking."[49] Thomson might have put aside these difficult and unrewarding issues had Tait not persevered.

Thomson and Tait disclaimed any novelty in their textbook:

> We desire it to be remarked that in much of our work, where we may appear to have rashly and needlessly interfered with methods and systems of proof in the present day generally accepted, we take the position of Restorers, and not of Innovators.[50]

Their wish to appear conservative was substantiated in many parts of the treatise. In a chapter titled "Experience," in the first division of the book, they reviewed the different classes of mathematical theories about forces "thoroughly known, [where] the mathematical theory is absolutely true, and requires only analysis to work out its remotest details." An example of this, the highest form, was the theory of planetary motion. The means of learning about astronomical phenomena were necessarily observational, whereas "in our laboratories, [when] we interfere arbitrarily with the causes or circumstances of a phenomenon, we are said to *experiment.*"[51]

The second class of mathematical theory was "based to a certain extent on experiment. . . . " An example given was the dynamical theory of heat which was "based upon the experimental fact that *heat is motion.* . . . " The equations derived from that "experimental fact" were "at present obscure and uninterpretable, because we do not know *what* is moving or *how* it moves." These theories had to remain in the limbo of partially verified theories until something was known of the "ultimate, or *molecular,* constitution of the bodies, or groups of molecules, at present known to us only in the aggregate."[52]

The third class of mathematical theories had the slimmest foundation in experiment. Examples of this class of theory were the mathematical theory of the conduction of heat (Fourier), of static electricity, and of permanent magnetism. In spite of the fact that no one knew *what* static electricity or permanent magnetism were or how they produced effects, "the laws of their forces are as certainly known as that of Gravitation, and can therefore like it be developed to their consequences, by the application of Mathematical Analysis." By working with effects rather than ultimate causes, Fourier, Green, Poisson, and Ampère produced useful examples of this type of mathematical theory.[53] The hierarchy in theory was the same as that taught to Thomson when he was a student at Glasgow. The major difference in the treatise's argument was in the validity of the use of mathematical analysis, the infinitesimal calculus, in deriving true relationships even when ultimate causes were not known.

Mathematics was an alternative to theory when it was impossible to learn as much as was needed, as in the case of mechanics, through experiment. The mathematics in the treatise was not an exercise in the "merely *curious*." They said that "in the present work we are engaged specially with those questions which best illustrate physical principles—neither seeking, nor avoiding, difficulties of a purely mathematical kind."[54]

Newton's laws were the foundation of the treatise, but Newton interpreted by means of that "grand principle of the *Conservation of Energy*," a principle, the authors declared, established by Joule's experimental work. They found that Newton anticipated the modern generalization "so far as the state of experimental science in his time permitted him. . . ."[55] The treatise, under the heading "Preliminary Notions," began with the purely abstract treatment of kinematics based on geometrical ideas. The second chapter of this section, "Dynamical Laws and Principles," used only three nonexperimentally established axioms, that is, Newton's three laws of motion. Starting with these well-founded preliminary notions, the treatise turned to its main study of dynamics.

The introduction to this second long, and as it happened, last division began by saying: "Until we know thoroughly the nature of matter and the forces which produce its motions, it will be utterly impossible to submit to mathematical reasoning the *exact* conditions of any physical question." Approximations based on the conviction that some information may be ignored without appreciably affecting the results were the best approach. The authors stated that the whole subject of abstract dynamics was to be examined with two principles in mind: first, the impossibility of the complete solution of any physical question, and second, that practical questions can be attacked by limiting their generality, *"the limitations introduced being themselves deduced from experience,* and being therefore Nature's own solution. . . ."[56]

While writing the treatise, Thomson and Tait discussed the basis for axioms. They were obvious only to those who had experience through experiment. There was no a priori or intuitive reason for accepting Newton's laws of motion. There was no apparent reason why circular motion might not be natural rather than motion in a straight line. In the printed version, this idea appeared as an introductory remark to Newton's three laws: " . . . it being remembered that, as the properties of matter *might* have been such as to render a totally different set of laws axiomatic, these laws must be considered as resting on convictions drawn from observation and experiment, *not* on intuitive perception."[57]

In making a distinction between what was intuitive or accepted a priori and what was Nature's own solution, the authors were led to question some of Newton's presuppositions. Immediately following the statement of Newton's first law were the qualifying remarks: "The meaning of the term *Rest,* in physical science, cannot be absolutely defined, inasmuch as absolute rest nowhere exists in nature." The first part of Newton's first law which states, *"Every body continues in its state of rest or of uniform motion in a straight line, . . . "* the treatise explained, merely expressed the convention for measuring time. "And the numerical measurement of time practically rests on defining *equal intervals of time, as times during which the earth turns through equal angles."* Where does this bring the reader? To the conclusion that the first part of Newton's first law is "of course, a mere convention, and not a law of nature. . . . " In the original draft, Thomson had written: "The meaning of the term ["rest"] in physical science has (we regret to say) not yet been explained." In a note to Tait, he added: "We indeed are not going to be so insane as to explain it; seeing that no such thing as a criterion of absolute rest exists in nature."[58]

The experience of writing the treatise had been fruitful, but the collaboration was too much for both men. The subsequent volumes which they planned were never published, but with the help of George Darwin, part two was published in 1883. Thomson said that by working alone, they could work on a greater variety of subjects more conveniently and not be restricted by the constraint of joint effort.[59] Soon, other physicists came out with textbooks modeled on the great treatise. Maxwell's *Treatise on Electricity and Magnetism,* for example, filled one of the voids left by Thomson and Tait.

The preface to the *Treatise on Natural Philosophy* promised a smaller version which would require only elementary geometry and algebra for the reader to understand the discussion. The smaller work, *The Elements of Natural Philosophy,* which was designed for the preuniversity student, appeared in 1872. The treatise was translated into German in 1871 by Helmholtz and G. Wertheim.

The first printing of the treatise sold out very quickly, but the buyers were not the general public. The intellectual community admired but did not read the book. The *Athenaeum,* "a Journal of Literature, Science, and the Fine Arts," was one of the few literary journals to take notice of the publication of the treatise. Their review in the October 1867 issue follows:

> A professor of Glasgow and one of Edinburgh, both well known to the world of science, have produced 750 [*sic*] pages as a first volume only. We defer description until we see the whole. The mathematical part is of

a formidable character, and of the most modern type. The authors are thoroughly up to their subject and have strong physical as well as mathematical tastes. The size of the volume is the fault of the subject and not of the authors, who have, so far as we have looked closely, kept down details. If anything, they have not sufficiently diluted the mathematical part with expanded demonstration. But what of that? The higher class students for whom this work is intended are rats who can gnaw through anything: though even *their* teeth will be tried here and there, we can tell them.[60]

Maxwell reviewed the second edition of the treatise for *Nature*. By this time, the second phase of the impact of the book had become apparent. Mathematicians have made important contributions to the study of dynamics, but only recently, Maxwell noted, was it possible "to open any memoir on a physical subject in order to see that these dynamical theorems have been dragged out of the sanctuary of profound mathematics in which they lay so long enshrined, and have been set to do all kinds of work, easy as well as difficult, throughout the whole range of physical science." Maxwell then said:

> The credit of breaking up the monopoly of the great masters of the spell, and making all their charms familiar to our ears as household words, belongs in great measure to Thomson and Tait. The two northern wizards were the first who, without compunction or dread, uttered in their mother tongue the true and proper names of those dynamical concepts which the magicians of old were wont to invoke only by the aid of muttered symbols and inarticulate equations. And now the feeblest among us can repeat the words of power and take part in dynamical discussions which but a few years ago we should have left for our betters.[61]

The *Treatise on Natural Philosophy* did more than any other book of the nineteenth century to make physics a specialized study. It organized the study of physics on the foundation of the new concepts of the nineteenth century. Until the idea of the dynamical approach to physical phenomena and that of the conservation of energy became common knowledge, the treatise would be a closed book to all but the specialist. Given the presuppositions of the two authors, Tait's wish to write a popular book was unrealizable. But not for John Tyndall, whose book *Heat considered as a Mode of Motion* (1863) was enormously popular. A review of the book in the *British Quarterly Review* pointed out the popular view of the difference between Tyndall's and Thomson's approach:

For anything like an exposition of the modern, and, we imagine, the true doctrine on Heat, we cannot do better than refer our readers to Professor Tyndall's pages. If the last degree of lucidity of treatment is not a matter of concern to them, Professor Thomson will be found perhaps an equally sure guide, though he may not conduct them quite so far. . . .[62]

CHAPTER TEN

The Limits of Science

The root of Thomson's disagreements with Tyndall and Huxley was based on their conflicting views of science. For Thomson there were limits to science, but for Tyndall and Huxley there were no limits. That boundless view led them to become the populace's representatives of science, a state of affairs that rankled the physicists who wore the mantle of inheritors of the tradition of Newton. The schism between the physicists and the popularizers became wider with the publication of Darwin's *Origin of Species,* in 1859.

In the 1860s, in the midst of his multifarious activities, Thomson the physicist became involved in a controversy with geologists over the theory of evolution and with biologists, especially Thomas Henry Huxley, the popularizer and reformer. The basis of the controversy was the question of geological temperatures and the age and previous condition of the earth, an early concern of Thomson's. Fourier's theory of terrestrial temperatures and the figure of the earth as interpreted by Thomson are keys to the early history of the earth.

Thomson's interest in the thermal history of the earth began in 1844, when the twenty-year-old Cambridge student wrote a paper titled "Note on Some Points in the Theory of Heat,"[1] in which he explored the mathematical issue about what happened when Fourier's equations had negative values of time. That is, was it mathematically possible to calculate the distribution of heat? In the equations, the initial distribution of heat was part of the data; it had to be given. Usually the equation was then used to determine the distribution of heat at some time *after* the initial moment. What happened if the equation was used to calculate the distribution of heat *before* the initial moment? Under some conditions, he found there was a solution.

Thomson pursued the subject further in his inaugural dissertation when he assumed the chair of natural philosophy at Glasgow. "De Distributione Caloris per Corpus Terrae" was never published, but the subject became fixed in his research repertory.[2] In an 1852 paper, Thomson pursued the implications for man of the second law of thermodynamics. The axiom was concerned with the availability of energy and, it said in effect, once the temperature of a body fell below a minimum, the energy in that body was not available for use, although the energy was not annihilated. Only a creative power, Thomson said, can make or annihilate energy. He concluded that there is a universal tendency in the material world for energy to be dissipated. Since that tendency has been going on in the earth and continues to operate, there must have been a time when the earth was too hot for man's habitation, and there must come a time when the earth will be too cold for human life. Unless, he wrote, some operation has occurred or will occur that is contrary to natural law.[3]

In 1854, in an article titled "On the Mechanical Energies of the Solar System,"[4] Thomson applied Joule's idea of the mechanical equivalent of heat to the sun in order to investigate the source of the sun's heat and the age of the sun. Thomson, in an astounding show of scientific daring, reviewed theories about the source of the sun's heat and found these reduced to three conceivable possibilities: that the sun was a heated body losing heat, that the heat is due to chemical action, or that the heat came from meteors falling into the sun in which case the motion of the meteors was converted into heat on impact. The meteor theory was suggested by Joule and adopted by Thomson for making his calculations as to the amount of heat produced and to estimate the age of the sun.[5]

To be certain that the three theories exhausted the possibilities was a highly speculative venture, indicating that Thomson was so convinced of the correctness of the physical theory that he confidently followed its implications. He believed, apparently, that an explanation of the sun's heat was possible on the basis of what was already known and did not pause to consider that there might be another cause not yet known. By this time, the dynamical theory of energy was a basic part of his thinking. Part of the theory was its universality, thus it was as applicable to the sun as to a steam engine. Here is one boundary line for Thomson's scientific thinking. In choosing one of the three theories, he was acting as if no important information existed about the sun that was not already known and that it was safe to establish a theory.

When can the mechanical reasoning of science legitimately be applied? Given the mechanical principles that explain the source of heat, the time can be determined when the sun began to heat the earth. By carrying the calculations forward, it was possible to establish the time when it was *"mechanically inevitable"* that the earth would no longer be habitable. In his paper "On the Mechanical Antecedents of Motion, Heat, and Light,"[6] Thomson introduced another boundary line for his scientific thinking. He believed that the moment of creation established the limit to mechanical speculation. Anything before that moment of creation was unknowable; that moment was the temporal boundary for science. When Thomson read the paper to the French Academy of Sciences, he was criticized by Abbé Moigno in his journal *Cosmos* because Thomson's ideas were outside the reigning doctrines of France and opposed the tradition of Laplace—or so the abbé thought.[7]

Thomson, who established himself as an international figure in science, that is, with a European as well as a British reputation, was the butt of the abbé's scientific chauvinism. At the beginning of the review in *Cosmos,* Abbé Moigno inserted the gratuitous remark that Thomson was the "spoiled child (I use this word intentionally) of English science."[8] The abbé was wrong in one respect. Thomson's ideas on the age of the earth annoyed some British scientists as much as they did the abbé.

In his paper, Thomson elaborated on his idea that the moment of creation established the limit to mechanical speculation. Mechanical reasoning, he said, supplied information that left no doubt that the earth was tenantless at some time in the past. Physical conditions were such that life was impossible. In addition to his certainty that matter and energy were created, he believed that all life was created. Purely mechanical reasoning, he said, "teaches us that our own bodies, as well as all living plants and animals, and all fossil organic remains, are organized forms of matter to which science can point no antecedent except the Will of a Creator, a truth amply confirmed by the evidence of geological history."[9] These thoughts, it must be remembered, were expressed sometime after Charles Lyell's publication of *Principles of Geology* (1830–1833) but before Charles Darwin's *Origin of Species* (1859).

Once impressed with these limitations to speculation about the future and past condition of the universe, Thomson thought a scientist was free to conjecture about the events within those time bounds. Next, Thomson speculated about the initial condition of the universe. Supposing there were, he said, a system of solid bodies at rest and at a great distance from each other, subject to only the force of gravity? These bodies, after a series of collisions, would achieve a state of motion, temperature, and light similar to our solar system and the rest of the universe. That was one possible explanation for the evolution of the universe to its present condition. Thomson concluded that if that theory seemed reasonable, gravity might in reality be the *"ultimate created antecedent"* of all energy in the universe.[10]

The cosmological picture drawn by Thomson was bracketed by the moment of creation and the moment in which all activity stopped. This would happen because of the lack of available energy in the distant future. Between these two points lay the vast span of time in which natural law operated. The dynamic world of constant change was diminishing in intensity as the amount of available energy declined. The further back in time, the more extreme the actions of nature. Thomson's position placed him in opposition to the uniformitarian geologists who assumed a constancy of the forces of nature, these forces being no more violent, they claimed, in antiquity than the present.

In his paper titled "On the Secular Cooling of the Earth," which Thomson read to the Royal Society of Edinburgh in 1862,[11] he said that geologists who opposed hypotheses using paroxysmal changes (the uniformitarians) overlooked the essential principles of thermodynamics. With the sun considerably hotter in past ages, according to the thermodynamic theory of the dissipation of energy, it was probable that there were extreme temperatures on earth in the distant past with more violent storms, and with vegetation of a more tropical type. In spite of his objections to the uniformitarian school, objections that preoccupied him for over eighteen years, Thomson proposed a compromise. Those geologists who suggested that catastrophes destroyed all life on earth were as counter to natural law as the uniformitarians. Thomson thought the probable explanation lay somewhere between the extremes of catastrophism and uniformitarianism.[12]

Thomson insisted that geological theory must be in agreement with sound mathematical theory and warned that those geologists who assumed a continuum of time—stretching back to infinity—began with an erroneous assumption. The true picture was of an initial condition on earth. That epoch could not have been the result of any previous condition of matter by natural processes. Thomson said that epoch was "well called an '*arbitrary* initial distribution of heat,' in Fourier's great mathematical poem, because that which is rigorously expressed by the mathematical formula could only be realised by action of a power able to modify the laws of dead matter."[13] That which was rigorously expressed mathematically must not be doubted. Geological speculation was therefore in error.

With the increasing popularity of geology in Britain during the 1860s, Thomson undertook to refute what he considered to be rash speculations that contravened mathematically based theory. He did this despite his problems with submarine cables, the treatise, and his classes. In another paper read in 1862, this time to the Royal Society, he began with a blunt statement: "That the earth cannot, as many geologists suppose, be a liquid mass enclosed in only a thin shell of solidified matter, is demonstrated by the phenomena of precession and nutation."[14] While geologists were basing their theory on strati-graphical observations, Thomson based his argument on the mathematical analysis of the rotation of the earth. When that analysis was in agreement with the observations of the earth's precession and nutation, it was, from Thomson's viewpoint, irrefutable.

The onslaught of Thomson's argument with the geologists continued in his publications in learned journals and popular magazines. In an article titled "On the Age of the Sun's Heat," published in *Macmillan's Magazine* in March 1862, Thomson had an opportunity to reappraise his earlier estimates. This article was a response to the popular reaction to his dire predictions. He thought that his earlier reckoning of the sun's future did not warrant pessimism. True, the second law of thermodynamics involving "a certain principle of *irreversible action in nature*" indicated an inevitable state of universal rest and death. But that terrible forecast assumed that the universe was finite and eternally subject to existing laws. It was impossible to conceive of a limit to matter in the universe, and reason led one to believe in "an endless progress, through an endless space. . . . " Regardless of what scientific theory was accepted, "it [was] also impossible to conceive either the beginning or the continuance of life, without an overruling creative power," Thomson wrote. "Therefore, no conclusions of dynamical science regarding the future condition of the earth can be held to give dispiriting views as to the destiny of the race of intelligent beings by which it is at present inhabited."[15]

What was a reader to think of Thomson's changeable speculations? In 1854, he estimated the sun's age at 32,000 years and predicted that its heat could not last more than 300,000 years.[16] He now thought it probable that the sun had not illuminated the earth for 100,000,000 years and was certain that the sun was not older than 500,000,000 years. "As for the future," Thomson wrote, "we may say, with equal certainty, that inhabitants of the earth cannot continue to enjoy the light and heat essential to their life, for many million years longer, unless souces now unknown to us are prepared in the great storehouse of creation."[17]

Why was Thomson, a foremost scientist of the day, who had made important contributions to the fundamental mathematical theory of natural philosophy, engaging in cosmological speculation? Was it a joke? Was he having fun with speculations about the future that were obviously disturbing to the uninitiated? Were his predictions merely a casual episode in his intellectual life? That hardly seems the case considering the number of papers he wrote on the subject and the amount of time and effort he gave to it. What of the seeming contradiction in his forecast for humanity? He spoke of there being no cause for dispiriting views of the destiny of man, but then he warned that inhabitants of the earth could not continue to enjoy the light and heat essential to the continuance of life. The prospect for man, in Thomson's cosmology, was that he was helpless but not hopeless. There is only one way to appreciate the seriousness of Thomson's prognostications, that is, by realizing that what he was saying was fundamental to his scientific beliefs and that his presumptions were basic to his private beliefs. The rigor and certainty of his research were made possible by limiting his questions. Fourier's equations gave the exact state of the temperature anywhere on earth, anytime in the future or the past, providing the necessary data were available and that the data included the initial distribution of heat for which there was no *physical* antecedent. On a larger scale, when the second law of thermodynamics prescribed the dissipation, not annihilation, of available energy, insights were gained into the physical condition of the physical world within the limits of scientific investigation. The possibility for scientific truth was unbounded within the limits of the law's competence. The universe was infinite, and the infinity of time and space was the realm of the Creator as well as the infirmity of man.

There was an intellectual movement growing in influence that viewed the object of science and its implications in a different way from Thomson. The science that became the source of the new way of thinking was evolutionary geology and biology. Their interpretation of the knowledge and theory of evolutionary geology, so recently gained, was that scientific knowledge was unlimited. The object of reform thinking was to promulgate the idea that man can know all and can control his own destiny. Appeals to an unseen and omnipotent diety, in their view, were futile and diverted man's attention from the task of improving his lot. Like its ancient predecessor, Lucretian atomicity, the new geology was observational and experimental. Belief in man's ability to observe and infer the ultimate cause of phenomena gives credence to the idea of unlimited possibilities for the human mind. In specific instances, prediction is extrapolated to the outer reaches of the universe and to the end of time. This view of the boundless power of the intellect contrasts with the mathematical theorist's assumption that rigorous truth is possible only by establishing bounds to man's quest for knowledge. Some of the more philosophical of the mathematical theorists did speculate beyond the bounds of what they determined to be the realm of certainty. That is the meaning of Thomson's and others' references to Fourier's "mathematical poem." The metaphor of that poem can be interpreted beyond the confines of the physically knowable universe when, and only when, one recognizes that the form of interpretation does not carry with it the degree of rigor which can be applied to limited physical questions.

William Hopkins, Thomson's Cambridge coach and a geologist, dissented from the views of the evolutionary geologists and biologists. He expressed opinions that Thomson must have agreed with; indeed Hopkins's influence can be detected in some of the ideas Thomson voiced on the limits of science. A mathematical theorist, Hopkins opposed Darwin's theory of evolution as well as many of the theories proposed by the evolutionary geologists. In an article published in *Fraser's Magazine,* Hopkins warned against an arrogance of intellect. The subject of man's biological history required that it be studied "not with that presumption which would claim even for some of the vaguest conclusions of man's intellect a weight and authority beyond that which may be assigned to higher sources of our knowledge." Like Thomson, Hopkins argued that biology was still at the stage of gathering evidence; evolutionary theory was not strong enough to hold the same degree of certitude as mathematical theories.[18] Hopkins had his own way of distinguishing between forces that were subject to the natural law and those without physical antecedent. In the case of forces without physical antecedent, the cause was a creative power, as Thomson would say. Hopkins said that the cause for some phenomena in nature was unknowable. For that reason, the assertion of an independent creation of species, which he believed in, constituted discontinuous physical causes. The idea of independent creation was discontinuous with known physical laws.[19]

Against criticism of evolutionary geology and biology as expressed by a few scientists such as Thomson and Hopkins was ranged the most effective defender of the new theories, Thomas Henry Huxley. Huxley was a popularizer of science, a supporter of Darwin's theory, and a scientist in his own right. He was elected a fellow of the Royal Society at the age of twenty-six and was a close friend of John Tyndall's.

Huxley's ideas about the objectives and limitations of science contrasted with Thomson's. Huxley's ambition was to promote the increase of natural knowledge and to forward the application of scientific methods of investigation to all the problems of life to the best of my ability, in the conviction which has grown with my growth and strengthened with my strength, that there is no alleviation for the sufferings of mankind except veracity of thought and of action, and the resolute facing of the world as it is when the garment of make-believe by which pious hands have hidden its uglier features is stripped off.[20]

Huxley's success as a popularizer through lectures and articles overshadowed the achievements of the less public theoretical scientists. Part of their antagonism toward Huxley was due to what they considered his misrepresentation of science and to their resentment of his popularity. It was said that "in England when people say 'science' they commonly mean an article by Professor Huxley in the *Nineteenth Century.*"[21] Huxley was an active protagonist of Darwin's theory because it was clearly a "resolute facing of the world as it is" and an effective force for staying those "pious hands." His reaction to *The Origin of Species* was immediate and delighted. Here was a scientific explanation to counter the biblical story of creation.

A major controversy over evolutionary geology and biology was conducted by William Thomson and Thomas Henry Huxley. The controversy between the two men was symbolic in that they represented two contrasting views of science. Thomson, the theoretical physicist, believed in the rigor of mathematics and the limitation of science. Huxley, representing Darwin, the observational scientist and the reform faction, saw the implications of scientific theory as an alternative to religious belief. Their debate centered on specific questions, but at issue were the objectives and methods of science.

At the end of 1865, Thomson read a one-paragraph paper to the Royal Society of Edinburgh called "The 'Doctrine of Uniformity' in Geology Briefly Refuted." Many of the "most eminent of British Geologists," Thomson said, held to the doctrine of uniformity, which assumed that the earth's surface and upper crust had remained unchanged in temperature and other physical qualities during millions of years. But at that time the heat given off by the earth, a quantity established by observation, was so great that in order for the earth to have maintained this rate of heat loss over 20,000 million years meant that the earth had to contain a very large quantity of heat in the beginning. The amount of heat lost from the earth and therefore presumed present 20,000 million years ago was enough to melt a mass of surface rock equal to the *whole earth's* bulk. No hypothesis that supposed chemical action, internal fluidity, effects of pressure at great depths, possible character of substances in the earth's interior, or any hypothesis "possessing the smallest vestige of probability," Thomson concluded, "can justify the supposition that the earth's upper crust has remained nearly as it is, while from the whole, or from any part, of the earth, so great a quantity of heat has been lost." Appended to his brief statement was an estimate of the present annual loss of heat from the earth using constants of conductivity for different substances recently determined by men in Edinburgh, Greenwich, and Sweden.[22]

Huxley complained that Thomson's estimates of the rate of cooling the earth were of the vaguest sort. "But is the earth nothing but a cooling mass, 'like a hot-water jar [References to illustrations used by Thomson] such as is used in carriages', or 'a globe of sandstone'? And has its cooling been uniform? An affirmative answer to both these questions seems . . . necessary to the validity of the calculations on which Sir W. Thomson lays so much stress."[23] Thomson's answer to both questions was yes. He believed that enough was known about thermodynamics so that a hypothesis could be chosen, and he rejected possibilities such as the idea of heat being supplied by undetected sources of chemical action that were heating the earth continually from within. He previously rejected the idea of chemical sources of heat, which were suggested by Charles Lyell. This hypothesis, Thomson said, was contrary to the principles of the dissipation of energy. (Lyell postulated a never-ending cycle of exchange of energy between light, heat, magnetism, electricity, and chemical affinity.) Thomson pointed out that Lyell's hypothesis was inconsistent with a number of facts dealing with conduction, radiation of heat, thermoelectric currents, chemical action and astronomy.[24]

He believed geologists were being led into error by repeated misinterpretations of the *principles* of natural philosophy. For example, the idea of the stability of the solar system had mistakenly been thought to mean the permanence of the solar system. The theories of the stability of the universe postulated by the French mathematicians Laplace and Lagrange omitted such disturbing factors as friction, which become of consequence when very long periods of time are considered. Laplace and Lagrange neglected these factors in order to simplify calculations. For example, tidal friction (the rubbing of the oceans against the ocean floor) must have slowed the earth's speed of rotation by half in about ten billion years (i.e., British billion which equals a million million). If the earth were solid so long ago, its shape would have become something much different. Yet some geologists would be satisfied with nothing less than that amount of time, ten billion years, for the earth to have been in a solid state. "Now," Thomson declared, "here is direct opposition between physical astronomy, and modern geology as represented by a very large, very influential, and, I may also add, in many respects, philosophical and sound body of geological investigators, constituting perhaps a majority of British geologists." The conclusion had to be "that a great mistake has been made—that British popular geology at the present time is in direct opposition to the principles of natural philosophy."[25]

Huxley used the occasion of his presidential address to the Geological Society of London in 1869 to reply to Thomson. As his text, Huxley took two sentences from Thomson's article "On Geological Time":

> A great reform in geological speculation seems now to have become necessary.
> It is quite certain that a great mistake has been made—that British popular geology at the present time is in direct opposition to the principles of Natural Philosophy.[26]

His main contention was that Thomson misrepresented geological theory. There were three schools of geological thought: the catastrophists, who believed that great forces such as the flood in Noah's time were needed to account for change; the uniformitarians, who insisted that it was permissible to postulate only the forces of change such as existed in the present—anything else entailed mystery and miracles; the evolutionists, whose theory embodied both the catastrophic and uniformitarian views. Thomson was wrong, Huxley contended, in saying that uniformitarianism was *the* theory of British geology because there were many young geologists who did not accept this view. Huxley believed that evolutionism would "swallow up" the other theories.[27]

Central to Thomson's objections was the idea of an infinite operation of physical laws, and it did not matter whether geologists advocated uniformitarianism or evolutionary theory, which was a combination of the infinite with the finite. It came to the same thing—avoiding the axiom which, he believed, brooked no modification: that time ran out on physical law. That axiom leads inevitably to the conclusion of the existence of a creative power.

That their differences were too fundamental to allow compromise can be seen in the perseverance with which Huxley sifted every one of Thomson's statements in search of error. Huxley contended that most of Thomson's points could be ignored because they were not definite. The discussion was "greatly embarrassed by the vagueness with which the assumed limit [of time] is, I will not say defined, but indicated [quoting Thomson] 'some such period of past time as one hundred million years.' Now does this mean that it may have been two, or three, or four hundred million years? Because this really makes all the difference."[28]

On the estimate of the age of the sun's heat, Huxley remarked that Thomson once held entirely different views as to the source of the heat. He was referring to Thomson's meteor theory. "But the fact that so eminent a physical philosopher," Huxley said, "has thus recently held views opposite to those which he now entertains, and that he confesses his own estimates to be 'very vague', justly entitles us to disregard those estimates if any distinct facts on our side go against them." When he did reply, Thomson said: "A British jury could not, I think, be easily persuaded to disregard my present estimate by being told that I have learned something in fifteen years."[29]

Thomson admitted to some uncertainties about the age of the earth and the age of the sun, but the change in the approximations was due to the inadequacy of the data and not to the failings of the principles of natural philosophy. If geologists required uniform conditions on the earth for so many millions of years, that assumed that the sun was giving off heat at a steady rate throughout this time. All things are possible to a creative power. To assume that the sun's heat was not waning would be the same as to suppose that the sun was a perpetual miracle, that is, that it contravened the laws of thermodynamics—laws that applied to all phenomena on earth. Such an assumption was inconsistent, Thomson cautioned, because geologists at the same time were insisting that the laws governing geological change were immutable.[30]

Thomson must have known that his criticism would cause a furor. If he was surprised by the number and intensity of the reactions, then he surely underestimated the popularity of geology and the adherence of professionals and nonprofessionals to certain fundamental hypotheses. Some of the outcry could be dismissed by Thomson with a smile. The *Pall Mall Gazette* proclaimed that geologists "on this side of the border" (i.e., England) were not yielding to the attack by a Scottish mathematician. In an article titled "Mathematics versus Geology," the *Gazette* complained: "There is in all this an element of uncertainty which utterly destroys the validity of Sir W. Thomson's evidence." The magazine presumed that "exact" mathematics was infallible. "Until the mathematicians agree, explaining the observed facts by the same hypothesis, it is hardly fair of Sir W. Thomson to ask geologists to accept his particular version as final."[31]

A. R. Wallace, who with Darwin formulated the theory of evolution, found it relatively easy to adjust the theory to Thomson's time limitations. In an article in *Nature*, Wallace suggested that evolutionary theory could be in agreement with Thomson's estimate of geological time. Perhaps, Wallace said, the effect of the precession of the equinoxes every 21,000 years was enough to produce extreme changes in temperature and thereby hasten the rate of extinction and modification of life on earth. He contended that this approach brought the development of the organic world to the period assigned by natural philosophy. Wallace saw that the controversy had little substance; both sides were fundamentally in agreement, and the differences could be erased by a modification of ideas which were not fundamental. Darwin had his mind fixed on a great expanse of time as necessary for evolution; Thomson's time limitations so upset Darwin that he likened Thomson to an "odious spectre."[32] But the shortening of the time span did not threaten evolutionary theory itself.

The *Edinburgh Review* surveyed the dispute in the article "Geological Theory in Britain," published in January 1870. They sided with Thomson in his attack on some of the loose speculations of the geologists as to boundless time in the past, but Thomson erred "in assuming that there is any necessary connexion between his limit of years and any geological theory whatever." It was known that "the eminent mathematicians and physicists" were not in agreement on tidal retardation, that the sun was merely a cooling body, or that something such as an aqueous atmosphere may have retarded the cooling of the earth in the past. With these points, Huxley carried "the war successfully into his opponent's camp." The article contended that "until all these questions are finally settled, it seems to us that any speculation as to the age of the earth based on purely mathematical considerations must be worthless. At present there are no data for their solution." The *Review* advised Thomson that "all such attempts to gauge the geological past by years can only end in vanity and vexation of spirit."[33]

Against Thomson's vexatious speculations was the issue of whether the uniformitarian theory was usable. That doctrine "has been mainly instrumental in raising geology to the rank which it now occupies among the sciences," and although important results had been obtained by investigations using the uniformitarian theory, "it seems to us that the time during which we have been able to observe existing phenomena is too short for a sweeping generalization. . . ."[34] It was contended that in order to be a science, geology was dependent on a modified uniformitarian theory of evolution. The dispute was between geological theory—one might say law—and speculation by the physicists.

Thomson was not prepared to agree with this view. The disagreements between natural philosophers on such matters as tidal retardation and whether the sun was merely a cooling body did not damage the validity of the laws of thermodynamics. Disagreements caused variations in the estimate of time, but there were enough cross checks on the physical theory, such as the rigidity of the earth and the calculations of the energy emitted by the sun, for there to be confidence in the certainty of the general principles. Thomson saw a primary difference between evolutionary geology and natural philosophy. Physical laws, and especially the thermodynamic laws, were established beyond doubt by experimental evidence and the logic of mathematics. These laws were certain and universal, but their very certainty was based on their being applicable over a definitely limited time span. The evolutionary geologists, Thomson said, had erred "in assuming that we must consider present laws—a very small part of which we have been able to observe—as sufficient samples of the perennial laws regulating the whole universe in all time."[35] No laws known to man were perennial!

The sensitivity of the geologists and biologists was founded on ample evidence that physicists accepted their theory with reservations and believed that geology and biology were not sciences—yet. In 1864, the Royal Society awarded Darwin the Copley Medal. The medal was a signal honor for Darwin and, through him, for all biologists. General Edward Sabine, president of the Royal Society, said of *The Origin of Species:* "Speaking generally and collectively, we have not included it in our award" to Darwin. He pointed out that while some agreed with Darwin's theory, "others may perhaps incline to refuse, or at least to remit it to a future time, when increased knowledge shall afford stronger grounds for its ultimate acceptance or rejection."[36]

In his presidential address to the British Association in 1871, Thomson left no doubt that he did not accept Darwin's theory. He quoted Darwin's concluding statement in *The Origin of Species,* but purposely omitted the sentences describing the origin of species by natural selection "because I have always felt that this hypothesis does not contain the true theory of evolution, if evolution there has been, in biology." It was not that Thomson subscribed to the Fundamentalist interpretation of the biblical Genesis. His response to that interpretation was, "If a probable solution [to the origin of life on earth], consistent with the ordinary course of nature, can be found, we must not invoke an abnormal act of Creative Power."[37]

Thomson rejected the theory of evolution through natural selection because he believed that the theory depended on accepting some explanation of the origin of the first living cell. "Dead matter cannot become living without coming under the influence of matter previously alive." Thomson's explanation for the origin of life on earth was the transportation of life here by meteors. "The hypothesis that . . . life . . . originated on this Earth," he said, "through moss-grown fragments from the ruins of another world may seem wild and visionary; all I maintain is that it is not unscientific. . . . "[38]

Thomson's scientific imagination was effective because he restricted it to prescribed limits. Beyond the reaches of man's grasp of physical law, he believed, was the creative power. What he sought in scientific endeavor, and what he believed about man's place in the universe was brought out in sharp outline during the controversy over evolution. The test of his private belief came when his wife, Lady Margaret Thomson, died on 17 June 1870 after seventeen years of searching fruitlessly for a cure—the years that encompassed their marriage. Despite her debilitating illness, they had forged a close-knit life. As much as she depended on him, he needed her.

Helmholtz described Margaret Thomson as a "charming and intellectual lady." She was also a poet. In 1866, a collection of her poems and German poetry that she translated was privately published as a surprise for her. The poems ranged from 1849 to 1869. S. P. Thompson said the following about her poems: "All are beautiful and some very sad: a note of sadness and death run through them." The volume was reprinted in 1874.[39]

A little more than a month after her death, Thomson wrote to Helmholtz:

> It is indeed . . . an unspeakable great loss which I have suffered. My sense of it goes on increasing every day and through all my occupations, after the first shock. That the end was certainly a happy relief for my dear wife from incessant suffering has done nothing yet to diminish the desolation in which it has left me.[40]

CHAPTER ELEVEN

Making the Metaphor Explicit

Three months after the death of his wife, William Thomson sought change in his life. He had enjoyed some trips on friends' yachts and associated sailing with the busy, challenging times spent on the *Agamemnon* and the *Great Eastern* during his Atlantic cable days. On those ships, he was free of land cares and able to do some mathematical work. So it was that in September 1870 Thomson bought a schooner, the *Lalla Rookh*. She was a 126-ton sailing ship built of oak and seventeen years old when Thomson became her owner. The advice he got assured him the ship was a "very good model and . . . a fast sailer."[1]

His first problem was how to outfit the *Lalla Rookh*. Mrs. Tait was his consultant and buyer. The question arose as to whether cotton or linen would be best for the *Lalla Rookh*'s berths. Thomson sought the advice of naval experts who said he should use linen rather than cotton because "cotton fabric seems to be too hygrometric to be suitable for sea-going places." In addition, he was concerned about the size of the sheets (he wanted them to be long enough for Tait) and the quality of the tablecloths (he wanted them to be of the highest quality damask in order to withstand the wear and tear at sea).[2]

S. P. Thompson describes the times Thomson spent weekends on the yacht without guests:

> Rising early he would take a plunge, before breakfast, in the sea, swimming round the yacht, and in spite of his lame leg climb with agility on board by rope. When there were no observations or soundings to take he would sit for hours with green book and pencil in hand working at calculations and meditating over his problems; or he would pace the deck smoking a quiet cigar.[3]

But he was happier with company aboard and solicited his friends and family at every opportunity.

"Will you not give us a visit on board the L. R.?", Thomson wrote at the end of a letter to Stokes. "Do come and we shall be able to talk over things deliberately, besides going out to look for waves." Helmholtz was amused to find the forty-seven-year-old Thomson talking as if the theory of waves was a race between them. When Thomson had to go ashore for a while he warned: "Now, mind, Helmholtz, you're not to work at waves while I'm away."[4]

Many letters to Stokes about scientific questions were written aboard the *Lalla Rookh*. In 1872, Thomson wrote to Stokes from the yacht that he was sending a paper, "Contact Electricity," to be read at the Royal Society. The paper was written out twelve years before, only waiting "till I should be able again to work on the subject which is now the case."[5]

Having no engines, the yacht offered its challenges and opportunities. Stokes received a letter from Thomson's secretary stating that Thomson, his brother James, and Helmholtz were becalmed and spending their time determining minimum-wave velocity. Stokes was treated to a detailed description of how the three yachtsmen trailed a weighted fishing line over the side in order to observe waves. A fog offered the chance to test Thomson's sounding machine and depth gauge under adverse conditions. James wrote to a friend that William "was quite in earnest feeling his way along the bottom yesterday in the fog by taking soundings and comparing his results with soundings marked on the chart." As if to show there was no cause for alarm, William wrote a note on the letter, "in from 30 to 105 feet."[6]

That the ship had become part of Thomson's life is evident in a note he added to a letter to Stokes: "The College, Glasgow is my proper address. I oscillate between the Laboratory and the L. R. in weekly periods, pausing longest on latter end of range (? because?) better adapted for work." In answer to a suggestion by Tait that Thomson become principal at a college, Thomson replied that it "would not do for me at all nor I for it." He wanted "nothing but to get through what work I am fit for, quietly. My desk in the NPL [Natural Philosophy Laboratory] and the L. R. are the only places in the world for which I am fit." Then, ironically, he closed the letter in haste because he had to go to Newcastle for a few days and then to London.[7]

In the spring of 1871, Thomson asked Helmholtz to come to the meeting of the British Association in Edinburgh the early part of August and then sail with him to the West Highlands of Scotland. He hoped to have Huxley and Tyndall on board too, but they would be occupied after the meeting. When the meeting took place, William Thomson was chosen president of the British Association. His presidential address was scheduled for 2 August, but by July he still had not written a word, although he was aboard the *Lalla Rookh*. When he finally finished writing his speech, it was long and ranged from the need for accurate measurement to the introduction of life on earth by meteors.[8]

Helmholtz was unable to attend the meeting, but he did come to Scotland in the middle of August. He had not seen Thomson since the death of Mrs. Thomson. Before joining him on the yacht, Helmholtz had the opportunity to visit Glasgow University in its new quarters and saw Thomson's new house.[9] Helmholtz felt compassion for Thomson in his widowerhood and was depressed by the untidiness of Thomson's house. The portrait of Mrs. Thomson that he found there also saddened him. He felt that Thomson's life was desolate without her.[10] It no doubt was. When he went to London in June 1871, Thomson had given a party for his friends from the Cambridge University Musical Society. He wrote to his sister-in-law that it was a "strange reunion, like a return from the other world. . . . " Although he had looked forward to the reunion, he realized "it [could] never again be what it was, and it is too full of sadness for the present."[11]

His work bored him. Later that year, he complained that "the things I have been kept incessantly busy with are dull and uninteresting, except so far as getting through little by little what must be done is interesting."[12]

Helmholtz and Thomson spent about a month on the *Lalla Rookh* sailing about Scotland. During the early part of September, they sailed for Roshven, the home of Hugh Blackburn and his wife, Jemima (Maxwell's cousin). Helmholtz wrote to his wife that Thomson was "very eager" to arrive there. Roshven was "a lonely property, a very lovely spot on a bay between the loneliest mountains." It was there that Jemima Blackburn found the birds that she drew for her book. "She fashions all her doings and household ways to suit her professional tastes. . . . It was all very friendly and unconstrained."[13] Jemima Blackburn was a lovely, talented lady and Hugh was Thomson's best friend. No wonder William Thomson was "very eager" to be in their company.

Thomson busied himself with many matters, especially those concerning the cable. In June 1873, Thomson was on the *Hooper*, a cable ship with 2,500 miles of cable, heading for Lisbon. On 29 June, Fleeming Jenkin wrote to his wife: "Here we are, off Madeira at seven o'clock in the morning. Thomson has been sounding with his special toy [a sounding machine] ever since half-past three (1,087 fathoms of water)." Due to a fault in the cable, they stayed in Madeira for more than two weeks. In that time, life began to brighten for William Thomson. The most important person in Madeira, Charles R. Blandy, Esq., made Sir William feel welcome at his villa, as did his three daughters. Thomson amused them by teaching them the Morse code.[14] But the stay in Madeira ended too soon. S. P. Thompson quoted an eyewitness report of how one daughter, Fanny Blandy, said good-bye:

> A figure was seen waving a floating streak of white drapery from a window of the house on the hill high above the port. "G-O-O-D-B-Y-E" was spelled out. "Eh! What's that? What's that?" said Sir William, adjusting his eyeglass the better to catch the signals. "Good-bye, good-bye, Sir William Thomson." And as the ship's hull dipped beyond the horizon the white streak still fluttered "Good-bye."[15]

Less than a year later, in March 1874, William Thomson announced that he intended to sail for Madeira on 2 May. Before embarking in his yacht, he left instructions to have the temperature in his house made the same as that in Madeira. As the *Lalla Rookh* approached Madeira Harbor, the story goes, Thomson signaled to the house on the hill, "Will you marry me?", and Fanny Blandy signaled back, "Yes."[16]

They were married on 24 June 1874, two days before Thomson's fiftieth birthday. (Fanny was thirteen years younger than he.) Thomson wrote to Helmholtz: "My present happiness is due to a fault in the cable which kept the *Hooper* for sixteen days in Funchal Bay last summer."[17]

In a way, it was fitting that William Thomson married so close to his fiftieth birthday, a time of change in many people's lives. After his marriage to Fanny, he began to do things in a grand way, and for the first time in his life, he became interested in the particulars of everyday living. His first marriage was filled with the illness of Margaret Thomson. The brash young man of seventeen who did not mince words in attacking the man (Kelland) who attacked his hero (Fourier) was restrained, at first by his father, but more permanently by the burden of his marriage. Consequently, a shadow was cast on his ebullience. With his marriage to Fanny Blandy, who was as healthy and active as he, the restraint was removed and the shadow lifted. Soon after their marriage, they built a house near Glasgow in the Scottish baronial style and called it Netherhall (for which Maxwell supplied peacocks). Tait thought the Thomsons' "palatial style" of entertaining was in marked contrast to his own.[18]

The man who spent almost his whole life in his father's house finally built a home of his own. Because of the baronial style of Netherhall, his mode of living took on a different cast. The way of life at Netherhall mirrored a personality long hidden from view as, in an analogous way, his increasing involvement in practical applications of scientific theory revealed another side of Thomson's work, that is, his dualistic approach to science, part theoretical and part practical.

The *Lalla Rookh* was one manifestation of Thomson's attachment to the sea. The pleasure the yacht afforded was amplified by the opportunities Thomson had for indulging his interest in oceanography. The theory of navigation became a major interest, and in addition to practicing the art on his yacht, he gave talks and wrote articles on the subject. One such talk, "Navigation," was given under the auspices of the Glasgow Science Lectures Association in the city hall on November 1875. That year, he also gave a talk on "The Tides" to 2,000 people at the same hall.[19]

In the course of preparing a popular article on terrestrial magnetism and the compass for the magazine *Good Words,* Thomson found that the compasses he examined were poorly designed. To compensate for the unmanageable features of their compasses, ships' captains were in the habit of weighting the compass cards or increasing the friction of the pivot point (one captain filled the pivot with brick dust) in order to make the compass "less lively." Thomson saw that these interferences lessened the accuracy of the compass. He thought there was a need for a more responsive compass that was also steady in rough weather and adjustable, allowing one to compensate for the effect of an iron ship on the compass needle. These stipulations raised the question of how to deal with the erratic terrestrial magnetism through which the ship was passing.[20]

The determination of the amount of compass adjustment required by an iron ship moving through the earth's magnetic field was done mathematically by Thomson's friend Archibald Smith. But Thomson realized that as the ship moved through the earth's magnetic field, the effect of the ship's iron changed. Therefore, some means was needed to make corrections periodically on a ship's compass. To make these corrections, Thomson designed and patented a system of movable iron globes and a magnetic corrector mounted on the compass binnacle. In 1876, Thomson tested the new compass on his way to the United States aboard a Cunard steamer and again on his return trip. It worked to expectation, and he was ready to promote his invention. In 1877, he learned that Lord Rayleigh and his wife were to sail on the *Balmoral Castle*. On board were three of Thomson's compasses, which he asked Rayleigh to check. They were in the wheelhouse.[21]

The British admiralty refused to consider Thomson's compass. Not only was the device founded on principles counter to popular belief, but it seemed so fragile. A model that Thomson lent to the admiralty gathered dust under a table in the admiralty office. The captain who hid it was heard to comment: "Our quartermasters would put their fists through that card." Thomson wryly replied, "It seems he [the captain] was resolved to guard it against the possibility of such an accident."[22]

Thomson's compass design recalled his technique in the fabrication of the mirror galvanometer. The compass card was unusually small, about 10 inches in diameter. He did everything possible to reduce the weight of the card and its magnetic moment. Eight small compass needles were attached to the card, and they weighed only 54 grains. The paper of the card had been cut away to reduce weight and was braced with an aluminum rim. The total weight of the compass card was 170½ grains, or 1/17 of the weight of an ordinary 10-inch compass card.[23]

The fragile card was sealed safely under glass; it worked well in bad storms, even under the shock of the ships firing a broadside. As promised, the compass was easily adjusted for changes in the ship's magnetism. Thomson was soon collecting endorsements by satisfied navy captains. He carried on a campaign to have the admiralty adopt the compass, but this effort occupied only part of his time. He had too many interests, both business and scientific, for him to devote too much of his time to any one activity. James White manufactured the compass, and there were agents selling the compass in several cities. Eventually, the Thomson marine compass became standard equipment on British ships and was adopted by other navies.

In 1876, two years after their marriage, Sir William and Lady Thomson sailed for the United States. Thomson was invited to be one of the 125 foreign judges at the Centennial International Exhibition in Philadelphia.[24] Thomson arrived in Philadelphia ten days late. Before his arrival, he was elected president of Group 25 (judging instruments of precision, research, experiment, and illustration, including telegraphy and music). Joseph Henry, secretary of the Smithsonian Institution, presided until Thomson arrived. Under the United States banner, Thomson "saw and heard Elisha Gray's splendidly worked-out Electric Telephone," and "saw Edison's Automatic Telegraph delivering 1015 words in 57 seconds. . . . " Under the

Canadian banner, he heard Alexander Graham Bell's telephone.[25] In Thomson's report of the device, he wrote that Bell "achieved a result of transcendent scientific interest—the transmission of spoken words by electric currents through a telegraph wire."[26]

The Thomsons were enthusiastic visitors. Lady Thomson wrote to Mrs. Tait: "We are becoming fearfully yankee. In fact we guess we have forgotten British." They saw Niagara Falls and visited Boston, Newport, Toronto, and Montreal. Thomson was "vividly impressed" with what he saw "both in the Great Exhibition in Philadelphia and out of it, showing the truest scientific spirit and devotion, the originality, the inventiveness, the patient persevering thoroughness of work, the appreciativeness, and the generous open-mindedness and sympathy, from which the great things of science come."[27]

They went to Washington and visited Joseph Henry, who was not only a "splendid old fellow" but also a "generous rival of Faraday in electromagnetic discovery." Thomson met Simon Newcomb in Joseph Henry's drawing room. Newcomb was a mathematician and astronomer who was then at the Naval Observatory in Washington. Thomson thought Newcomb was a "first rate man—full of *go*, and seems to have excellent capacity. This time 50 years, by the time mathematics gets fairly engrained into the Americans they will have left us fairly behind I suspect. They are immensely zealous, and eager to go ahead in scientific science, as well as practical inventions of all kinds in which they are now beating us, rapidly, particularly in land telegraphy . . . the public profits greatly by the simplicity and excellence of their patent laws."[28]

On Thomson's return to Glasgow, he carried his praise of American science to the Mathematics and Physics Section of the meeting of the British Association in Glasgow.[29] The *New York Times* reprinted Thomson's remarks under the heading "Science in America." Undoubtedly, this was done as an antidote to Simon Newcomb's long sober assessment of American science in an article titled "Abstract Science in America 1776–1876," which appeared in the January 1876 issue of the *North American Review*.[30] According to Newcomb's view, American science had no place to go but up. However, anyone reading Thomson's remarks had to believe American science had already arrived.

On his return from America, Thomson became even more involved in technology. Helmholtz worried about the waste of Thomson's talents. He thought Thomson might better devote his time to scientific research than to designing instruments which Helmholtz thought were too subtle and delicate for most workmen. "He is simultaneously revolving deep theoretical projects in his mind, but has no leisure to work them out quietly." But Helmholtz admitted, "As far as that goes, I am not much better off! . . . I did Thomson an injustice in supposing him to be wholly immersed in technical work; he was full of speculations as to the original properties of bodies, some of which were very difficult to follow; and, as you know," he wrote to his wife, "he will not stop for meals or any other consideration."[31]

In part, Thomson's continued interest in scientific research was a consequence of his work in technology, specifically the technology of measurement, which opened the way for his later work in the less mathematical and more physical theory of the ultimate nature of matter. Following the lead of Fourier, Green, and Poisson, his earlier mathematical papers avoided physical hypotheses about the nature of heat, electricity, or magnetism. Physical causes being unknown in "the present state of science,"[32] Thomson used mathematical analogy for developing theories about the different forms of energy. Joule's idea of convertibility and equivalence between different forms of energy, the dynamical theory, became in Thomson's hands the mathematical metaphor whereby the different forms of energy were related through mathematical abstraction. Still there was a barrier to progress toward an understanding of these phenomena *physically.* Thomson's equations describing the force exerted by terrestrial magnetism were not physical theories since they did not answer the question, What caused terrestrial magnetism? One of the paths that led Thomson to these fundamental questions was his involvement in the attempt to define absolute standards of measurement. Another path converging on the theory of the ultimate nature of matter and energy was his practical work in inventing systems of measuring the tides and determining magnetic bearing at sea.

At Thomson's suggestion, the committee on Electrical Standards was appointed by the British Association in 1861. Besides Thomson and Fleeming Jenkin, four others were appointed to the committee. But Thomson was unhappy. He wrote to Jenkin that *"all and sundry"* had been appointed. "When I saw Wheatstone and Williamson had been named, I thought Siemens and Sir C. Bright, and any other practical man who had expressed an interest in the finding of units might still more appropriately be added. I cannot think that in such a committee Siemens could give much trouble." Jenkin replied that he doubted the addition of Siemens, Bright, or even of Latimer Clark, an engineer, could mitigate his fear of getting only an "illusory report. It will be next to impossible to get these men to agree to any plan of action. There are only you and myself who cordially support the Absolute unit." In 1862, Siemens, Bright, Maxwell, and two others were added to the committee.[33] Thomson, Jenkin, and Maxwell, together with James Prescott Joule, who was appointed later, were the mainstay of the committee during its eight years of labor.

The committee's first goal was to establish units for measuring electrical phenomena for both practical and scientific work. Recognizing that no single standard that they would derive could satisfy two sets of criteria that in no way overlapped, they determined to work solely on an absolute unit of measurement.[34] Practical units had to be convenient, of a convenient size for instruments, and simple in definition so that technicians could understand the basis for units. All aspects of the practical system of units were dictated by the industry that used them. The submarine cables required that certain characteristics be measured, such as resistance and inductance, in the magnitudes encountered in those long lines and to an exactness commensurate with the relatively small telegraph signal currents. As the electric-lighting industry developed, new quantities, such as power consumed in watts, needed to be measured; the magnitudes of the quantities measured were a good deal larger than those for the telegraph. Absolute measure was closely allied with theory, and the units were a necessary aid in the quantitative description of the theory. Units of measurement served the dual purpose of making the theory more explicit, thus making it experimentally confirmable.[35]

In a lecture he gave to the Institution of Civil Engineers in 1883 on "Electrical Units of Measurement," Thomson talked about the need to advance beyond reference to a set of numbered standards. He insisted that "more is necessary to complete the science of measurement in any department; and that is the fixing on something absolutely definite as the unit of reckoning" before the subject, in this case electricity and magnetism, was accessible to scientists. He might have added experimental scientists. By the end of the talk, the engineers should have understood that Thomson associated the absolute system of measurement with scientific theory and not engineering. Perhaps they were misled by his opening statement where he used that often quoted remark: "I often say that when you can measure what you are speaking about and express it in numbers you know something about it. . . . " He neglected to add, but believed, that one knows one thing about the phenomena when it is measured in practical units, but a different idea is suggested when using absolute units. The system he described had no direct relevance to engineering. In Thomson's view, "When you cannot measure it, when you cannot express it in numbers, your knowledge is of a meagre and unsatisfactory kind: it may be the beginning of knowledge, but you have scarcely, in your thoughts, advanced to the stage of *science.*... "[36]

There was an echo there of the words spoken by Dr. Nichol to Thomson's natural philosophy class in 1840. Nichol said: "The object of science is the discovery of relations . . . of which the complex may be deduced from the simple. We must fix on some common attribute to compare objects. . . . Cant [Kant] has shown that all may be ranged under the head of quantity. Now no branch of physical science is perfect until it can be reduced to number or quantity."[37]

When Thomson devised his absolute thermometric scale in 1848, he used Carnot's motive power of heat. Thermometric measurement was only as accurate as the consistency of the substance used in the thermometer over the range of temperatures measured, and since no substance was thoroughly invariable, the measuring of temperatures was at best a system of arbitrary points of reference, which was accurate enough for practical purposes, but not for scientific investigation.[38] But Thomson's absolute temperature scale was independent of particular properties of substances and was based on the mathematically precise relationship between temperature difference and mechanical effect produced, as expressed in Carnot's theory.

The establishment of absolute units of measurement went hand in hand with the advancement of theory. Absolute units were built from a base of the best established theory, mechanics—the science of motion. In the nineteenth century, heat, through Thomson's efforts, was translated into absolute terms. After the connections between forms of energy were established by the dynamical theory, electricity and magnetism were ready for their turn. It was a step-by-step foray from the known to the less known, measuring each step of the way.

Thomson referred to absolute units as a universal means of measurement, but recognized that ideally they were not universal, although the units were as universal as they could possibly be. In his notes he wrote: "It may be objected that it is not universal because from Jupiter or Venus or distant stars it would not be easy to observe the Earth's angular motion. Let the inhabitants of those bodies appear and make their complaint. Till then we shall regard the absolute unit as a universal or sufficiently so, unit." These musings were serious in their intent, since Thomson recognized that the second of time, a basic unit for the absolute system, was derived from the motion of the earth. Earth time was not universal in the strictest sense, but it was sufficient for the scientists on this planet. Thomson compared the absolute system with the medieval scholars' use of Latin as a universal language. He thought the use of the idea of force and work expressed in the absolute system had the same advantage that Latin had in the Middle Ages. But, he declared, the absolute system had to be translated into terms understandable to intelligent mechanics of Glasgow.[39]

Thomson thought the practical man should be nurtured and encouraged. His displeasure was evident when the Royal Society rejected a paper of Latimer Clark's for the society's *Transactions*. Thomson considered the paper to be an original, important contribution to absolute measurement. He wrote to Stokes that the rejection of the paper was a "great discouragement to practical men and an injury to science if when they do anything so exceedingly good as this in the way of raising the scientific character of their profession, the Royal Society should throw cold water on their efforts." If he thought the referees would ignore "the great scientific importance of absolute measurement," he would have had the paper transmitted by someone else on the chance he would be chosen one of the referees.[40]

Thomson found that Joule's work provided valuable information for the absolute system in extending the concept of mechanical effect to more and more actions involving the conversion of energy. By stating these conversions in terms of mechanical effect, it was possible to return always to the base of the system—mechanics. Joule had experimentally determined the mechanical equivalent of the chemical action of a battery; the concept was then extended to chemistry via the battery, then to electricity. And by measuring the heat produced in a wire, the concept was extended again. By the addition of the mechanical effect produced by the electromagnetic action of a current flowing through a wire, the general principle of mechanical effect through all these steps of reasoning became the basis for an absolute system of units in electricity and magnetism.[41] The absolute system would be as useful as Latin and would serve the scientists as a conceptual scheme very much the way pointillism was an imaginative device for the impressionist painters of the nineteenth century. The means of expression was an avenue for the imagination. The absolute system bound together the different forms of energy in a concrete way and operated on the imagination of scientists in a more direct way than the abstract metaphor of the mathematical equations.

Thomson's advocacy of the absolute system was influential first in the work of the British Association committee and later in international conferences that adopted most of the British Association units. Gauss and Weber in Germany did important work for the absolute system of electricity, but Thomson's efforts were behind the adoption of the system. That the British Association Committee on Electrical Standards interpreted their charge as being the search for the absolute unit of measurement indicated the extent of Thomson's influence. The committee's report for 1863 took as its starting point an electrical unit of measurement that could be expressed in terms of length, time, and mass. The report stated, "In fine, the word "absolute" is intended to convey the idea that the natural connexion between one kind of magnitude and another has been attended to, and that all the units form part of a coherent system." They contended that the absolute system was the only rational one. The committee's stated object was—and if the words were not Thomson's, certainly the thought was—that "in determining the unit of electrical resistance and the other electrical units, we must simply follow the natural relation existing between the various electrical quantities, and between these and the fundamental units of time, mass, and space."[42] Once the world community of scientists adopted the absolute system of electrical measurement, the mathematical theory of electricity and magnetism was made physically explicit.

The basis for the whole system was the definition of the absolute unit of electrical resistance. The British Association committee derived a definition by imagining a conductor moving through the earth's magnetic field. Making use of Faraday's discovery of electromagnetic induction and Oersted's discovery of the magnetic effect of an electric current,[43] the imaginary circuit enabled the committee to define the absolute unit of resistance as 1,000,000,000 centimeters per second. Thus the committee used theory to develop a concrete concept expressed in absolute and universal units. Thomson admitted that stating resistance in centimeters per second "conveys a strange idea," but "it is perfectly true as to the absolutely definite meaning of resistance." In addition, the committee suggested several methods of measuring resistance that would be useful to practical men, for example, by means of a column of mercury. Thus the dual objective was satisfied: the absolute system was nearly universal, and the practical system was convenient. As far as Thomson was concerned, "The first step toward numerical reckoning of properties of matter" had been made.[44]

In 1874, Thomson became president of the Society of Telegraph Engineers. In his presidential address, he tried to interest the group in participating in a worldwide endeavor of studying atmospheric electricity. "In these great changes of terrestrial magnetism, or in the mere existence of terrestrial magnetism, we have always before us one of the greatest mysteries of science—a mystery which I might almost say is to myself a subject of daily contemplation. What can be the cause of this magnetism in the interior of the earth?" He thought the subject of atmospheric electricity had a close connection to the telegraph—there was the theoretical connection between earth currents, which plagued telegraphists, terrestrial magnetism, and perhaps the aurora borealis. Thomson proposed a simultaneous worldwide reading of atmospheric electric charge; readings were to be coordinated through the telegraph.[45]

Thomson was a one-man enterprise in urging and doing whatever possible in carrying out his plan for the global study of atmospheric electricity, which he called "electrogeodesy." He put his students to work taking readings with instruments mounted in a classroom and from the Glasgow University tower. He sought ways to have his electrometer used. He hoped to interest a balloonist in taking a portable electrometer aloft with him and offered to instruct him in taking readings. The possibility of taking readings in an inaccessible part of the globe by an Arctic expedition led Thomson to offer not only to supply the expedition with Thomson electrometers but also to oversee their production and test them himself, if only they were used for taking a few observations. He sent an electrometer to Nova Scotia, and, at the Smithsonian Institution, Joseph Henry ordered one.[46] At the 1859 meeting of the British Association, Thomson gave a talk titled "On the Necessity for incessant Reading, and for Simultaneous Observations in different Localities, to Investigate Atmospheric Electricity."[47]

Atmospheric electricity was a long-standing interest of his. In the summer of 1856, Thomson and his first wife, Margaret, had gone to Krueznach for her health. During their stay, Thomson met Friedrich Dellmann who besides being *Oberlehrer* in the gymnasium was in charge of meteorological and electrical observations for the Prussian government. Thomson thought that Dellmann's electrometer was a great improvement on earlier ones. It left "nothing to desire in the way of accuracy, as far as regards meteorological requirements." Thomson had an electrometer made to take back to Glasgow.[48]

In a Friday evening lecture titled "On Atmospheric Electricity," given at the Royal Institution in May 1860, Thomson described the improved electrometers he invented. He sent a copy of his lecture to Dellmann, who thanked Thomson "for the zeal with which you used the hint I gave you during your visit in the summer of 1856."[49]

Others were not as grateful for Thomson's enthusiasm for their ideas. In his private notebook, Tait wrote: "W. Thomson strikes me far more as a man of extreme readiness to improve upon the *slightest* hint, than as an absolutely original thinker. That he displays extraordinary ability in doing so is incontestable. But I have found so many of my *own* new ideas developed by him (sometimes immediately, sometimes after the lapse of a year or two) that I think he must have done much the same to others." Tait attributed Thomson's actions to "the eccentricities of a *most absolutely honest* man. There is no doubt that he cannot distinguish between what he thinks, and what he hears:—for he *never* pays full attention to anything, he is always *also* thinking on something else; so that what he hears gets mixed with what he thinks; and he takes it for his own."[50]

In 1895, Lord Rayleigh told a demeaning story about Thomson that corroborated Tait's view. But in 1870 Tait took a different view. He wrote to Thomson: "Don't you be a fool and let yourself be pumped by our friend. *Publish* what you please, but don't write long private letters except to people who are likely to make an honest use of them."[51]

A number of people might complain of Thomson's unacknowledged use of their ideas, but many more had good reason to be grateful for the generous way Thomson gave them credit. Edward Roberts was annoyed that Thomson used his ideas, and he managed to provoke Thomson. Roberts was either naive or, more likely, he was in a position to gain something by embarrassing a public figure.

In 1879, Thomson heard that Edward Roberts, a clerk in the Nautical Almanac Office, was calling the Thomson Tide-Predicting Machine the Roberts Tide-Predicting Machine. Roberts accused Thomson of appropriating the key idea for the machine from him; therefore, the tide-predictor was his and not Thomson's. Unlike Thomson's younger days, when he dealt ineffectively with Whitehouse and the Scottish universities commission, he dealt forthrightly with Roberts.

The episode began in 1867 when Thomson asked the British Association to conduct a study that would not only supply navigational information, but also contribute to a better understanding of the forces that acted upon the earth's surface to cause tidal motion. The British Association then appointed a committee "For the Purpose of Promoting the Extension, Improvement, and Harmonic Analysis of Tidal Observations," with Thomson as its chairman. In his memorandum to the committee, he asked them to consider the need for

tidal reading at regular intervals of time rather than just at the two
irregular occurrences of high and low tide. According to his dynamical theory of tides, the many forces acting upon the oceans were
never at rest. Thomson thought that tidal theory was "one strong
link in the grand philosophic chain of the Newtonian theory of
gravitation. In explaining the tide-generating force we are brought
face to face with some of the subtleties, and with some of the mere
elements of physical astronomy."[52]

To calculate the height of tide water at *any* moment required the
solution of a mathematical equation consisting of a number of
simple harmonic functions. The components of tidal force were described by Thomson as the lunar semidiurnal tide, the solar semidiurnal tide, the lunar diurnal tide, and solar diurnal tide. There
were, besides, the lunar tide occuring every two weeks and the solar
semiannual tide. The diurnal and semidiurnal tidal forces were not
uniform because of the eccentricity of the moon's and earth's orbit;
the semidiurnal tidal forces were irregular because of the varying
declinations of the sun and the moon. To determine the level of
water at any moment in a harbor meant that equations involving all
these components must be solved simultaneously. Thomson wanted
the height of water recorded every three hours for a month in order
to supply the data needed to make tide tables. He also wanted a
greater degree of accuracy than had hitherto been obtained. He
wrote to his brother James that the readings should be correct to
within one hundredth of a foot. Since Thomson knew of no gauges
with this degree of accuracy, he designed one himself. He gave
James a detailed description of the new tidal gauge and asked him to
have one installed in Belfast Harbor. Thomson planned to place one
in Glasgow Harbor.[53]

In 1875, Thomson petitioned the British government to extend
the tidal observations in the Indian seas. Not only would the additional readings correct the insufficiency of the British admiralty's
tide tables for Indian ports, Thomson wrote, but the year-to-year
observations by means of tidal gauges would be also desirable for
scientific purposes. The admiralty could take advantage of the British Association Tidal Committee's work, which had thus far cost
£1,000. The committee had been at work since 1867 using Thomson's method of analysis, and, he reminded the Indian council, these
results were "thoroughly available for practical application." A further argument for taking the necessary number of readings was that
a tide-predicting machine would soon be completed and available
for public use at the South Kensington Museum. Thus, if harmonic

analysis data were available for a port, the machine could produce tidal information for years in advance.[54]

The method developed by Thomson for analysis and prediction of tides was derived from Laplace's work in *Mécanique Céleste* in which the several forces acting on the earth's waters were considered at the same time and from the point of view of constant change. Time, then, became one of the variables in the equations that took into account the changing positions of the earth, moon, and sun to each other. All these forces operated simultaneously on the earth. In this study of the tides, a number of Thomson's earliest interests were brought together. Laplace's book and the notion that the earth was a rotating ellipsoid covered with water recall Thomson's "Essay on the Figure of the Earth." In addition, Thomson used Fourier's analysis in order to simplify the complicated curve that resulted when plotting the combination of all the tidal forces. All these mathematical operations required long and tedious hand calculations.

As chairman of the British Association Tidal Committee, Thomson hired Edward Roberts to oversee the lengthy computations. Roberts devised a system that used teams of clerks in India and some student help from Thomson's natural philosophy class. Repetitious calculations of this type indicated the need for a calculating machine, or as Thomson said, it was a place "to substitute brass for brain." At this point in the development of the system for calculating tides, Thomson had the collaboration of his brother James whose contribution to the tide-predicting system was an idea for the harmonic analyzer, a machine he had used for meteorological calculations. The key element was a modification of a planimeter along lines suggested to James by James Clerk Maxwell.[55]

The mechanization of the third stage was missing. The first stage was the collection of accurate and complete data with the tidal gauge. These data would then be put through the harmonic analyzer to find the constituent forces acting on the water in any given harbor. Then these results had to be calculated by Roberts's crew of clerks. In 1872, Roberts visited the *Lalla Rookh* to discuss some tidal calculations with Thomson who talked about building a calculator to do the work. They were unable to devise anything satisfactory. The principle of the calculator was known, and different mechanisms that performed similar operations had been built. Some of these were described in Thomson's and Tait's *Treatise on Natural Philosophy*, but none of them suited their purposes. After the meeting with Roberts, Thomson continued to work on the problem as he traveled to the meeting of the British Association in Brighton. Beauchamp

Tower, his traveling companion, suggested using the chain and pulley mechanism incorporated in Wheatstone's ABC letter telegraph instrument. "I had known this arrangement," Thomson recalled later, "but when Tower explained it to me I jumped at it and instantly knew it was the very thing for me." Before the end of the journey, Thomson had completed his design.[56]

The basis of Roberts's claim as inventor of the Tide-Predicting Machine was a different version of the above story, that is, he and not Thomson acted "on the suggestion of the chain passing over pulleys." And it was he, unknown to Thomson, who instructed Légé (a London instrument maker) on how to make the successful machine. Later, in public, he strongly disagreed with Thomson for saying otherwise.[57]

Roberts's duplicity came to light when Thomson offered to transmit a paper by Roberts on the tide-predictor to the Royal Society. Thomson wanted "to give him an opportunity of coming before the Royal Society as having had the superintendence of the construction of the tide-predicting machine for the India Office." Upon learning the title that Roberts had put to the paper, "Preliminary Note on a new Tide-Predict[o]r," "which seemed to imply that the instrument was a new one" invented by Roberts, a correspondence ensued between Thomson and Roberts that was filled with such anger that Stokes, not Thomson, transmitted the paper.[58]

Thomson found himself in a difficult position. One would think that a man of Thomson's stature could have easily quashed Roberts. But there are always those who are ready to believe that a person with great accomplishments has come by them in a not wholly honest way. So it was that Thomson, not Roberts, had to defend his reputation. In July 1879, Thomson wrote to Stokes explaining how the situation with Roberts occurred:

> My great difficulty with him [Roberts] and Légé has been to have my own ideas carried out, and I have often regretted very much that I did not have the whole done by White in Glasgow—with his genius to help, the details would have been much more satisfactorily carried out—and I should have had *infinitely* less trouble. . . .
>
> Roberts has been most excellent and useful in all the tidal calculations, and I left wholly to him the determination of the numbers of teeth . . . [for the wheels] and to him and Légé the planning of the shafts to bring the motions to the cranks in the positions in which I placed them; but every other detail I fought out *against them* in each case in which it was novel. They did their best but their dynamical intelligence was not superior to that of ordinary instrument makers and, on that account, not from any wish on their part to obstruct me, and because they were 400 miles

from my home, I had infinitely more trouble and a less satisfactory result than I should have had with White.[59]

The affair dragged on for almost two years and culminated in three meetings, held in March 1881, of the Institution of Civil Engineers, where it was finally laid to rest.[60]

Priority in invention, just as priority in scientific ideas, is not assigned easily because there are a number of steps in the realization of an idea. First of all, there is the original concept, followed by synthesizing the concept with other ideas to conceive of a complete mechanism or theory, and finally, there is the indispensable step that makes the idea workable and acceptable. Patent suits attempt to assign priority to one person, and priority disputes in science aim misleadingly for the same result—choosing a single person to be credited with the theory. The tide-predictor controversy placed Thomson in a position similar to other nineteenth-century innovators, notably Thomas A. Edison: he was the person who invented the central idea of a machine manufactured successfully by others.

In his business affairs, Thomson preferred an ad hoc arrangement for managing each of his inventions. In this way, he could direct his own organization and be in control. At one time or another, the Thomson engineering enterprises consisted of three separate partnerships: James White Optician (later Kelvin and James White, Ltd.); for telegraph work, Jenkin, Varley, and Thomson; and Jenkin and Thomson. Other than these three separate agreements, there was no overall organization, such as Siemens Brothers had.

Thomson, Varley, and Jenkin was formed to protect patent rights on submarine cables. Jointly they controlled the important patents that any company needed in order to operate a cable. Cromwell F. Varley was little more than a business associate, but Fleeming Jenkin had a close personal and professional relationship with Thomson that began in 1859, when Jenkin was twenty-six and Thomson thirty-four. Jenkin had been trained as an engineer, but Thomson encouraged him to write about scientific subjects. Thomson nominated Jenkin to be a fellow of the Royal Society, and he was elected in 1865. Jenkin and Thomson entered into a joint patent in 1860 for telegraph signaling apparatus, and the agreement to work jointly with Varley was made in 1865. Jenkin managed most of the detailed work in his partnership with Thomson. In 1869, he was appointed to the chair of engineering at Edinburgh.[61] After 1875, Thomson was not involved in telegraph affairs, and in 1883 the partners dissolved their agreement.

The telegraph years were profitable for all three men, and since Thomson held the major patents he received the largest share. The business founded on patents had its own peculiar worries; their association no sooner began when they found it necessary to threaten suit against the Atlantic Telegraph Company. The company had tried to deduct a large part of their revenue, £125,000, as being exempt from royalty payments. The issue was settled out of court in 1869 with the Atlantic Telegraph Company paying a lump sum of £7,000 plus £2,500 per year for the next ten years for the use of all the patents held by the three men.[62]

In 1880, the partners were engaged in a legal battle with the Western and Brazilian Telegraph Company. After a trying time during which the company threatened a counter suit of £300,000, a settlement was reached. The arrears in patent fees was reduced from £7,250 to £4,000, and Jenkin and Thomson accepted an annual royalty of £2,000 instead of £3,000. Thomson asked Varley to also accept an abatement of his royalties. Thomson felt he could be generous in view of the handsome returns he was realizing on his patents. For example, his share from the partnerships in 1879 was over £5,500.[63]

In the 1880s, Thomson was absorbed in electric lighting. His measuring instruments were as applicable to the lighting industry as they had been to the telegraph. Many of Thomson's patented electrical instruments, such as the galvanometer, electrometer, and the resistance bridge were easily adapted for use in the new industry. The laboratory began to build and test new devices. James Thomson Bottomley, Thomson's nephew, was given charge of the electric lighting work.[64] (Another nephew, William Bottomley, Jr., was in charge of the nautical devices such as the sounding machine and the compass.) James Bottomley was sent to Paris to investigate the latest developments in lighting equipment. The *Lalla Rookh* was to have electric lights with the aid of the new Faure accumulators. "Sir William is *tremendously* interested in it," Lady Thomson wrote to George Darwin.[65] She often served as Thomson's secretary.

Frances ("Fanny") Thomson enjoyed and probably did the most to maintain the high level of activity at Netherhall. When there were no visitors, she lamented that the house was "so quiet." One of Thomson's great-nieces described the activities at the house during the late summer of 1892: There were twenty-four guests, and an endless number of others coming to lunch, tea, and dinner. She noted that her uncle did not seem to mind the company and went on with his mathematics in the midst of the turmoil. Occasionally he talked animatedly with whomever was nearby. The niece relished one quiet period when there were no guests at all in the house for a total of thirty hours.[66]

Thomson maintained his professorship at Glasgow; he had no further academic aspirations. The atmosphere at Glasgow was best for him, and he refused all offers, including the Cavendish professorship. In 1871, Cambridge University established a professorship in experimental physics in conjunction with the Cavendish Laboratory for teaching heat, light, electricity, and magnetism. Thus, belatedly, the university sought to correct their neglect of physics. Thomson was first choice for the position at Cambridge. Maxwell wanted him to accept the offer because he had more experience in teaching experimental work. He refused and Maxwell was finally chosen as the first Cavendish professor.[67] When he died in 1879, Thomson was approached again.

Maxwell's death, at the age of forty-nine, was a blow to science. When Thomson heard of Maxwell's illness, he wrote to Stokes: "I can scarcely bring my self to think we are to lose him so soon, with so much of pleasant intercourse as we hoped to have, and so much to do for science before we all go."[68] Maxwell had started their association twenty-five years earlier by writing to Thomson: "As I wish to study the growth of ideas as well as the calculation of forces, and as I suspect from the various statements of yours that you must have acquired your views by means of certain conceptions which I have found great help, I will set down for you the confessions of an electrical freshman."[69]

Perhaps Thomson should have accepted the position at Cambridge. Things would have been quieter, and he might have finished more of the work he had pending. Often the engineering *did* interfere with his scientific work. In a letter to E. J. Routh in 1881, he apologized for neglecting to answer a letter: "I have had scarcely a corner of my brain available for mathematics. . . . All this time I have been so dreadfully pressed (chiefly with electric lighting, which I have now done all through my house from attic to cellar). . . . " But Thomson had found time, and he enclosed some mathematical work that Routh might find interesting.[70]

Writing anything of any length was agony for Thomson. In 1879, he was commissioned to write an article on heat for the ninth edition of the *Encyclopaedia Britannica*. He often complained how long it was taking him. He wrote to Stokes that the article was past due. Although he had spent every possible moment on it, it was still only one third done. When the new edition of the *Britannica* was reviewed by the *Nation* in 1882, the reviewer noted that Thomson's article "Heat" had a "ludicrous omission"—Tyndall's name was not mentioned once.[71]

In 1879, when the *Treatise on Natural Philosophy* was revised and became part one, and in 1883, when part two was published, Thomson was fortunate in having the dedicated services of George Darwin, Charles Darwin's son. He undertook much of the responsibility for the revisions, making many trips to Netherhall. Without him, the revisions would not have been done.

Given Thomson's restless and dynamic nature, it might appear at first that the time he spent with inventing and engineering was time out of his more important work in science. But by a process of unconscious selection, he had become engaged during this period in a number of technological enterprises that were tests of his basic scientific theories developed during his youth. His work on absolute measurement, the compass, the tides, and other enterprises were a materialization of the abstract. Once he had made explicit the metaphors of his theory, he could venture further into consideration of the fundamental nature of matter and energy.

CHAPTER TWELVE

The Finger-Post to Ultimate Explanation

In 1884, at the age of sixty, William Thomson gave a series of lectures, later called the Baltimore Lectures, at Johns Hopkins University. President Daniel C. Gilman of Johns Hopkins invited Thomson as part of an effort to make the university a leading center for science in the United States. Arthur Cayley, a mathematician from Cambridge University, had given the previous series. Wolcott Gibbs of Harvard first suggested to Gilman that he invite Thomson, but they "hardly thought it likely" that Sir William Thomson would accept.[1]

American science was in its adolescence. American scientists knew the British were doing more important work than they, as well as the reason why.[2] They were overwhelmed by Sir William Thomson's contributions to science and held him in awe. The friendly, ebullient Scotsman put them at ease. Besides, he was seen as a father figure who was at least twenty years older than most of them. Wolcott Gibbs was an exception, he was two years older than Thomson.

Thomson learned from Gilman that the "nucleus" of the audience would be "a dozen of our Associates, Fellows, and advanced students, who have been well trained in Mathematics and Physics." One of the first questions to be settled was the subject of the lectures. Wolcott Gibbs wrote to Gilman that "Sir Wm. Thomson will choose his own subject," but Thomson gave them a choice: vortex motion, gyrostatics, molecular dynamics, Fourier mathematics, and equilibrium of an elastic solid. He was willing to speak on two or three of the subjects he listed. Gilman sent Thomson the reaction of Wolcott Gibbs who was opposed to any of the usual topics such as mechanics or thermodynamics because they were of little interest to American scientists. He hoped Thomson would talk about the *"obscure* and *difficult* points in our modern physics." Among the difficult points that came to mind were the wave theory of light, the atomic and

202

molecular theory of matter, and electricity "as regards the want of any physical theory whatever." He hoped for "a really vigorous showing up of our shortcomings, especially if supported by new views such as Thomson could and would bring forward."[3]

Henry Rowland informed Thomson that the topic that would interest the greatest number of American scientists was molecular dynamics. His next choice was vortex motion. He concluded his letter with the statement that Thomson's lectures at Johns Hopkins would "always be considered an event in the history of science" in America.[4]

The reason for this trip, Thomson's second to America, was not only to give the series of lectures but also to take part in the meeting of the British Association in Montreal. Lady Thomson accompanied him. Their ship docked in New York, and they spent a few days with Cyrus Field and his wife at their country place in Irvington, New York, before traveling to Montreal for the meeting.[5]

The meeting in Montreal was the first to be held outside of Great Britain since its founding in 1831. When the decision to meet in Canada was announced, the Canadians chose to see it as recognition of their scientific stature. *Science* recognized the meeting as an opportunity that has "never before offered itself," to exchange ideas with scientists from Great Britain. The *Popular Science Monthly* called the British Association the "most powerful of scientific organizations" and recognized its presence in America as "the gathering strength of the great scientific movement in this age."[6]

It was estimated that 800–900 people came from Great Britain. The total number registered at the meeting was 1,773.[7] Among the Americans attending were Thomas Mendenhall, Henry Rowland, Simon Newcomb, and E. L. Youmans, editor of the *Popular Science Monthly*. The *New York Times* noted the presence of "such distinguished men as Sir William Thompson [*sic*]," but wondered at the absence of "great men like Tyndall and Huxley."[8] Undoubtedly, this comment caused more than one of the British scientists to gnash their teeth.

The Canadian government and the city of Montreal were so thrilled with the presence of such distinguished scientists that together they contributed $40,000 toward the expenses of the meeting, which was held at McGill University. In addition, the Common Council of Montreal held a reception where the mayor welcomed the group—a gesture that led the reporter of the *New York Times* to imagine what would happen if the same event happened on Manhattan Island. The thought presented an incongruous picture: "Fancy Mr. Alderman Kirk and his kind presenting an address to Sir William Thompson [*sic*]. . . ."[9]

But Sir William Thomson overcame the stereotyped picture of a titled Englishman and, in a disarming manner, introduced the incoming president of the British Association, Lord Rayleigh:[10]

> In the field of the discovery and demonstration of natural phenomena Lord Rayleigh has, above all others, enriched physical science by the application of mathematical analysis; and when I speak of mathematics you must not suppose mathematics to be harsh and crabbed. (Laughter). . . . I will not, however, be hard on those who insist that it is harsh and crabbed. In reading some of the pages of the greatest investigators of mathematics one is apt occasionally to become wearied, and I must confess that some of the pages of Lord Rayleigh's work have taxed me most severely . . . . [11]

When Thomson gave his presidential address, "Steps toward a Kinetic Theory of Matter,"[12] to the Mathematical and Physical section of the British Association, he indulged his penchant for chasing apparently irrelevant thoughts. Nevertheless, unlike those who were annoyed by that habit, *Science* was charmed and, it can be inferred, the other Americans reacted in a similar manner:

> The whole address was enriched by the most delightful digressions . . . during which the manuscript was neglected, and the section was afforded the pleasure of following as best it could, the great physicist in his involuntary excursions into this most interesting but little-explored domain of physical science.[13]

The American Association for the Advancement of Science (AAAS) had scheduled its annual meeting in Philadelphia to begin on 4 September 1884, a day after the meeting of the British Association closed. For the convenience of the visitors, a special train was scheduled to travel from Montreal to Philadelphia. About 300 of the Britains came to Philadelphia. Held concurrently with the meeting was an electrical exhibition at the Franklin Institute. The Britains were urged to see the exhibition because it showed "what has been done [in America to forward] the utilitarian applications of electricity."[14]

Thomson gave a paper at the AAAS meeting titled "On the Distribution of Potential in Conductors experiencing the Electromagnetic Effects discovered by Hall." *Science* noted the prominent part Thomson played at both the British Association and AAAS meetings, which was "of unusual interest and importance to the members of section B [physics]. . . . In particular the section was fortunate in having the presence of Sir William Thomson. . . . Whenever he entered any of the discussions, all were benefited by the clearness and suggestiveness of his remarks."[15]

The meeting was plagued by an intense heat that affected a good part of the East Coast in September and October. When the meeting ended on 12 September, Thomson and his wife left to spend two days in Boston and two in Cambridge, which they "enjoyed *very* much." As Lady Thomson wrote to George Darwin, "The only cool days we have had were at Southampton in Long Island, Sag Harbour and Shelter Island, on our way to Boston."[16]

On their way south to Baltimore, where Thomson was to give his lecture series, they had to stop in Philadelphia. Thomson had promised to give a popular lecture there titled "The Wave Theory of Light." It was given on 29 September at the Academy of Music.[17]

The next day, 30 September, they were in Baltimore, Maryland, at Johns Hopkins University. At the age of sixty, Thomson was still indefatigable. That evening, he was guest of honor at a reception, which was also attended by Henry Crew who kept a diary. At the time, Crew was a twenty-five-year-old student of Rowland's. Crew recorded that Thomson "made some remarks on our projected Physical Laboratory and on Harvard's new one—said it was the most complete and practical he had ever seen—then made some remarks on the undulatory theory of light—said he most thoroughly believed in the existence of a luminiferous ether."[18]

Thomson and his wife, Fanny, were happy to settle down in Baltimore for a while. But the heat persisted, and Lady Thomson wrote to George Darwin that "every thing is hot and dry and parched. The leaves of the trees are falling instead of colouring, and the perpetual blue sky is quite fatiguing."[19] On 1 October 1884, Thomson gave the first of his twenty lectures, "On Molecular Dynamics and the Wave Theory of Light." *Science* reported that

> Sir William Thomson's course of lectures at Johns Hopkins University has opened with every prospect of being a brilliant success. It would be difficult to find a case in which a lecturer on so abstruse a subject was greeted with so large and appreciative an audience as was collected in Baltimore to hear our distinguished visitor.[20]

The lectures gave Thomson an opportunity to work out his ideas in an exchange with many scientists. The impatient Thomson was unable to develop his ideas in a long treatise, but under the pressure of his talks, which he did not have organized beforehand, he formulated into a coherent entity speculations that had been on his mind for years. His listeners were forward-looking scientists, some of whom were about to make their reputations. For example, there was Henry A. Rowland, who was about to become one of the best-known

American physicists of the nineteenth century. His talent had already been recognized by Maxwell, who sponsored Rowland's first major paper for publication in the *Philosophical Magazine*. Rowland was thirty-six. Albert A. Michelson, a thirty-two-year-old naval officer assigned to audit the lectures, and Edward A. Morley who, in 1887, would conduct the experiment on the speed of light that was to challenge the ether theory. Morley was forty-six. Cleveland Abbe, a student of Walcott Gibbs and European trained, was applying research in atmospheric physics to the work of the Signal Corps' Weather Service. Abbe was thirty-five at the time of Thomson's lectures.[21] Some of the British scientists who attended the meeting in Montreal were also there. Lord Rayleigh, Cavendish Professor at Cambridge, had just given the presidential address at the meeting in Montreal. He was later to discover argon gas, for which he was awarded the Nobel Prize. Rayleigh was forty-one. Also from Britain was George Forbes, the son of Thomson's friend James D. Forbes, who had been professor of natural philosophy at Edinburgh. George Forbes was a scientist and engineer, and he was to design the electric generators for the Niagara Falls power plant after all others had failed. All were prepared for a confrontation between the established nineteenth-century science, as represented by Sir William Thomson, and the new generation of scientists. Fortunately, someone was there to take notes of the proceedings. Arthur S. Hathaway, a Johns Hopkins mathematician, acted as stenographer.[22]

"Molecular Dynamics and the Wave Theory of Light" was one of Thomson's efforts to attempt a *physical* explanation dealing with the ultimate nature of matter and energy. In order to be acceptable to him, such an explanation must explain all known phenomena. He rejected his previous attempts as failures when he discovered discrepancies. But he still believed that "the grand object is fully before us of finding a comprehensive dynamics of ether, electricity, and ponderable matter, which shall include electrostatic force, magnetostatic force, electromagnetism, electrochemistry, and the wave theory of light."[23]

The trial this time was by light. A medium for transmitting all forms of energy had to be multifarious in order to serve in many different situations. The medium Thomson had in mind was the ether, and he was going to divine what properties the ether needed in order to provide a medium for all the phenomena of light from reflection and refraction to the bending of polarized light in a magnetic field. Once the ether with the properties he attached to it met the test of light, he then had to show its appropriateness for all other phenomena.

The wave theory of light, as opposed to the particle theory, assumed that something, an ether, moved to cause the known effects. Thomson was certain that it was an elastic solid medium, whatever its other properties were. An elastic solid ether was a necessary starting point, he thought, in the description of the medium. By considering the ether to be an elastic solid, the major properties of light were accounted for. Thomson said that one of the stumbling blocks of the mathematical theories that assumed an ether was the presumption that the ether was continuous matter, and he thought the ether theory had foundered in trying to explain the dispersion of light experiments. How can light be broken into many beams unless it collided with discrete particles in the medium through which it passed?[24]

The mathematical theory of light was complete in many respects through the work of such mathematicians as MacCullagh and Cauchy, who had described reflection off of metallic surfaces mathematically. What was wanting, Thomson said, was a "dynamical explanation" of light.[25] Dynamical theories all began with the assumption that something must move in order to produce a force. What was the exact description of that something, the ether in this case?

A further stipulation of the dynamical theory was that what moved was molecular—imperceptibly small. A dynamical theory of light had to be in keeping with the dynamical theory of gases, which supposed minutely small gas particles, or the dynamical theory of heat, which posited heat as being due to the motion of molecules, or the dynamical theory of magnetism, which suggested that electric currents around minute particles caused magnetic attraction. But what was the nature of these smallest of particles and how did they move? What was it that moved to produce thermal effects? The dynamical theory of gases as developed by Maxwell was mathematically correct whether it was assumed that gases consisted of solid elastic balls or geometric points of force having no mass.

There were alternatives to the dynamical view in science. Not all scientists subscribed to the injunction that a physical explanation was a requisite of scientific theory. Some were satisfied with a mathematical view but, even though Thomson had developed a number of successful mathematical theories, he thought of physical explanations as necessary but temporary expedients on the way to ultimate explanations. For some scientists, Maxwell among them, the model was a satisfactory alternative to a physical explanation.[26] The model was an explanatory device. It was not purported to explain things as they were, but to imagine something that functioned as if it were real. That something, weightless fluids, elastic balls, or vortices of imponderable fluids, obeyed molar, mechanical, and mathematically describable laws in such a way that the model was made to produce the observed effects. By no means was the model supposed to be real.

Thomson began his third Baltimore lecture with advice to teachers and researchers to make models for themselves. Several times during the lectures, he suggested possibilities for models, showing the audience some crude examples. His suggestions were always tentative. Referring to one of the models he had made, Thomson said: "This is the simplest mechanical representation we can give of a molecule or an atom, imbedded in the luminiferous ether, unless we suppose the atom to be absolutely hard, which is out of the question." Moved by the spirit of participation, which Thomson encouraged, Rowland constructed a model to illustrate molecular vibrations. Thomson was delighted with the results.[27]

Pierre Duhem, the French philosopher and physicist, deplored the model as a pointless and confusing device of "the ample but weak mind of the English physicist." Duhem accepted many of Thomson's theories while ridiculing Thomson's use of the model. He found that Thomson's model had come after the discovery and was not necessary to the discovery. The model was used as a means of explanation. Nowhere has "the use of mechanical models shown that fruitfulness nowadays attributed so readily to it."[28] French mathematicians, who like himself were reared on strict mathematical logic, found them incomprehensible. Duhem found Thomson broader than most men trained in mathematical logic and viewed the Thomson-Maxwell school as distinct but puzzling:

> To a mathematician of the school of Laplace or Ampère, it would be absurd to give two distinct theoretical explanations for the same law, and to maintain that these two explanations are equally valid. To a physicist of the school of Thomson or Maxwell, there is no contradiction in the fact that the same law can be represented by two different models. Moreover, the complication thus introduced into science does not shock the Englishman at all; for him it adds the extra charm of variety. His imagination, being more powerful than ours, does not know our need for order and simplicity; it finds its way easily where we would lose ours.[29]

The scheme of the Baltimore lectures appears to be that Thomson was exploring the possibility that the model of the ether was a true representation. In his introductory lecture, he said: "It seems to me that there must be something in this molecular hypothesis, and that as a mechanical symbol, it is certainly not a mere hypothesis, but a reality." His approach to the question of the properties of the ether was daring, but he carefully based his speculation on experimental data. He refused to leave the ether as an unfathomable mystery declaring that more was known about the ether than the ultimate nature of air, water, glass, or iron. Furthermore, there was less to know about a substance as simple as ether. A great deal might be inferred about the ether from the well-established properties of light. In order for reflection and refraction of light to occur, the ether had to be absolutely incompressible; otherwise these effects would produce waves of condensation in the ether. The dispersion of light indicated that the ether consisted of particles. Thomson went so far as to calculate the mass of a cubic kilometer of the ether.[30]

One of the most troublesome theories abroad at the time, which, Thomson complained, upset the opportunity to make the dynamical theory a universal theory was Maxwell's electromagnetic theory of light. Thomson described it as the "so-called" electromagnetic theory of light. Maxwell's theory, Thomson pointed out, was based on the gross view of matter, and, being mathematically complete, the theory neither contributed to the ether theory nor needed that theory as a necessary assumption. A look at Maxwell's work,[31] Thomson suggested to his audience, showed that "Maxwell's electro-magnetic theory of light was essentially molar; and therefore not in touch with the dynamics of dispersion. . . . " Thomson thought Maxwell's theory was a paradox, for Maxwell was "one of the foremost leading molecularists of the nineteenth century. . . . " Evidence of Maxwell's contribution to the understanding of the ultimate nature of matter was his work in the kinetic theory of gases and his estimates of the size of atoms.[32]

Another aspect of Thomson's criticisms of Maxwell's theory came to the attention of George F. Fitzgerald, professor of natural and experimental philosophy at Trinity College, Dublin, when George Forbes wrote a review of the Baltimore lectures for *Nature*.[33] Forbes quoted at length from one of the 300 copies made of Hathaway's lecture notes. In a letter to *Nature*, Fitzgerald questioned the accuracy of some passages that were quoted on the subject of Maxwell's "so-called" electromagnetic theory of light. Thomson replied that he had indeed been quoted correctly and referred Fitzgerald to his article "Velocity of Electricity," appearing in Nichol's *Cyclopaedia* (1860).[34]

Fitzgerald then wrote a letter in which he went into detailed criticism of Thomson's ideas about Maxwell's theory. Fitzgerald sent the letter to Thomson and asked if he had any strong objections to what he had written. Thomson did have a few objections. Fitzgerald made the changes and asked Thomson to look it over again: "If there is not anything very bad in it and if you don't object I would feel very much obliged if you would forward it for publication in 'Nature'. . . . I need hardly say that my letter to 'Nature' is not intended to inform *you* of anything as I might as well teach my grandmother to break eggs but to correct what I fear will be an impression made by your words." In 1885, Fitzgerald's letter, with Thomson's additions, was printed in *Nature*.[35]

The Baltimore lectures were an unusual opportunity for Thomson. From the first lecture given at 5:00 P.M. on Wednesday 1 October until the twentieth at 5:00 P.M. on 17 October, his topic was the problems and achievements of nineteenth-century physics. Since he was a prime mover in the advancement of that science, Thomson might have assumed the role of apologist for what he called the "nineteenth-century school of plenum."[36] Both Faraday and Maxwell were dead, and Thomson acted as interpreter for their school of thought. His audience was eager to hear an argument in favor of a scientific point of view. They were competent enough for as philosophical an analysis as Thomson cared to give. They had purposely chosen what they thought was going to be a controversial series of lectures. But many were disappointed with Thomson's tentative espousal. Henry Crew, Rowland's student, wrote in his diary: "It was a grand failure—so say all the professors who understood his *math:* he did not put things clearly." *Science* commented that "the title 'Molecular Dynamics' does not give an accurate idea of Sir William Thomson's recent course of lectures. . . . It is impossible here to give any specific account of the lectures. . . . "[37]

The ideas that Thomson discussed were important to him and fundamental to his scientific outlook. He had never been combative, and in Baltimore, he refused to be Sir William Thomson, lecturing to skeptics, declaring what he knew to be right. He made no pronouncements and no firm statements that might be argued with. A reporter noted that Thomson's manner was "conversational" and that he "made almost no use of notes, and was full of enthusiasm for his subject." Thomson wanted to be "on a conferent footing" with the physicists present.[38] He called them his "coefficients." George Forbes, one of the coefficients, wrote that the "discussion did not end in the lecture-room, and the three weeks at Baltimore were like one long conference guided by the master mind. It is not surprising that at the end of that time there was a genuine feeling of sadness at parting on the part of teacher and taught alike."[39]

The Baltimore lectures were not an occasion for Thomson to make his testament as a scientist. He refused the opportunity. His lectures were not a carefully thought-out argument needed to carry the day for the plenum school. Much of what he said was as spontaneous as his lectures delivered at Glasgow. Rayleigh was amazed at Thomson's informality. He noticed that many of the morning lectures were based on questions that had been raised while the group was having breakfast.[40]

Were the lectures wholly exploratory? Thomson said at the beginning that his object was to examine the difficulties of the wave theory of light *"with the hope of solving them if possible;* but at all events with the certain assurance that there is an explanation of every difficulty though we may never succeed in finding it."[41] Was it possible that Thomson did not *want* to find the answer? He drew back from any final answer although he said time and again that the object of science was a final answer—a physical explanation. All of his life, his research was a pursuit without fruition. Thomson's mathematical papers always avoided physical hypotheses as being untenable in what he called "the present state of science." The appeal of the mathematical to his imagination was its abstractness—its indefinite and open characteristic. When he finally confronted the definite and had the chance to postulate the final physical answer, he refused to take the step. Science was the eternal quest. A final answer was an awesome thing from which he recoiled.

Thomson's persistent search for the ultimate, the physical explanation, was, he expected, interminable. In his presidential address to the Institution of Electrical Engineers, he said his question of the ultimate nature of matter was constantly in the back of his mind; it had been on his mind for forty-two years, and he recalled the precise moment when the idea occurred to him. It was in the first year of his professorship, in fact the first month, November 1846, when he wrote in his mathematical notebook that Faraday's experiment on electrostatic induction suggested the elastic solid analogy.[42] Inevitably, the analogy suggested a reality—that space was filled with an elastic solid substance. How could he persist for so long and so constantly without a measure of success?

At the 1884 meeting of the British Association in Montreal, when Thomson gave his presidential address to the Mathematical and Physical Section, he said the properties of matter were so connected that it was inconceivable that an explanation of one property of matter was attainable without having a theory that explained *all* the properties of matter. "But though this consummation may never be

211

reached by man," he said, "the progress of science may be, I believe will be, step by step towards it, on many different roads coverging towards it from all sides." A move along one of those roads, he thought, was the kinetic theory of gases.[43]

Thomson's second trip to America was a triumph, and his return to Glasgow was a flattering confirmation of his scientific eminence; he was asked for the third time to become the Cavendish professor. The chair of experimental physics at Cambridge became vacant when Lord Rayleigh, its second holder, resigned on his return from America. Thomson had no sooner started the session at Glasgow on his return from America when he received a petition from 167 resident members of the Senate of the University of Cambridge, asking him to become a candidate for Rayleigh's position. The petition was forwarded to Thomson by the master of Peterhouse at the behest of J. J. Thomson. Thomson declined. "My life's work is cut out for me here," he wrote to the master.[44] Thomson wanted Stokes to be a candidate, but Stokes refused. Lady Thomson wrote to George Darwin that Stokes had probably made a wise decision because "he is best plodding on by himself. . . . " Her choice was Tait.[45]

Richard T. Glazebrook, who had worked under Maxwell and Rayleigh at the Cavendish Laboratory, was an active candidate. However, when J. J. Thomson was chosen to succeed Rayleigh, Glazebrook was very disappointed. William Thomson was on the board of electors to fill the position. He wrote to George Darwin that he was "greatly grieved to think of Glazebrook. Lady R[ayleigh] writes to my wife that her husband is much upset about the appointment, to think of a person so little experienced in the laboratory work should be his successor. Still, I think the right thing was done. It seems to me most decidedly that the next best appointment would have been Glazebrook. But every one of us believed that J. J. Thomson is the stronger man; and what else could we do than what we did? If he has the capacity we believe him to have he will *very soon* make up for all want of experience in details of laboratory *management* and *teaching*. He is I believe as sure and trustworthy an experimenter, and is proved by what he has done to be so, as any of the other candidates."[46]

J. J. Thomson had won the Adams prize in 1882 with his essay "A Treatise on the Motion of Vortex Rings." It was published in 1883, and Thomson wrote to George Darwin that "I am becoming hot on vortex motion through having . . . J. J. T's book at hand."[47]

For a number of years, Thomson thought that another road was his theory of the vortex atom, a road that appeared for a time to lead directly to the consummation. In a paper titled "On Vortex Atoms" and delivered in 1867, Thomson gave Helmholtz's law of vortex motion in a perfect liquid as the source for his idea of a vortex atom. The perfect fluid in a peculiar motion, which Helmholtz called *Wirbelbewegung,* was not an analogy, not a model. It was, Thomson believed, the true atomic structure. It not only had the correct physical properties that would eventually explain all the phenomena, it was in motion. Here, then, was the ultimate substance necessary for the dynamical theory. The ultimate particle itself was matter in motion, and here the trail of science ended. "To generate or to destroy 'Wirbelbewegung' in a perfect fluid can only be an act of creative power," Thomson declared.[48]

The model for the vortex atom that helped imagine the atom and which led to an explanation of some of its characteristics was P. G. Tait's smoke rings. Tait had designed a box in which he generated smoke chemically and expelled perfect smoke rings by striking a diaphragm. He used it for demonstration in his classes, but Thomson saw the rings as a visible representation of the vortex atom. There was enough evidence for Thomson to see that this was the right path. The impenetrable atom of Lucretius had too few properties to begin to account for the variety of known phenomena. Thomson thought the vortex atom successfully accounted for the thermodynamic properties of gases revealed in recent experiments as well as being an explanation for the kinetic theory of gases.

Thomson illustrated the almost endless number of ways the vortex atom accounted for the varieties and allotropies of simple substances and their affinities for each other with diagrams and models. The vortex atom, Thomson claimed, had the necessary properties to account for Stokes's dynamical theory of spectrum analysis. The fundamental particle of matter, Stokes's theory claimed, was in a state of vibration, and the frequency of the vibration accounted for the different colors in the spectrum produced by heated elements. What better explanation for this vibratory motion, Thomson asked, than vortex atoms whose very nature was motion? Certainly, this explanation was superior to any suggestion that the massy Lucretian atoms vibrated.[49]

The conditions for vortex motion were mathematically describable, and Thomson was prepared to demonstrate the truth of that statement. The mathematical problem was soluble, and the whole situation presented mathematical difficulties of an exciting character. Time after time, Thomson returned to the idea of an ultimate, mechanical explanation for matter and energy. If it was an exaggeration for him to say it was on his mind day and night for forty-two years, it was true that he spoke frequently on the subject over a number of years. Time and again, he found a flaw in the idea; one thing he required of the ultimate explanation was that it be flawless—it had to explain everything without exception or it was good for nothing. In 1881, he reached another of those pauses in his quest (there was never a termination). The difficulty this time was trying to explain gravity and its relation to the inertia of masses.[50] Thomson's lament was: "No finger-post pointing towards a way that can possibly lead to a surmounting of this difficulty, or a turning of its flank, has been discovered, or imagined as discoverable." Nevertheless, no other approach suited, and the answer *had* to be in this direction.[51]

Thomson gradually abandoned the idea of the vortex atom; periodically he gave it up only to return some time later when new information seemed to make the idea tenable. He had little success in developing a following for the theory. In 1883, Tait wrote to Thomson that he had a means for destroying the vortex atom for good. However, that was before the 1884 Montreal meeting, and apparently Thomson was not dissuaded by Tait's proof. In his 1889 speech, Thomson once again examined all the advantages and difficulties of the plenum as an ultimate explanation and concluded: "But, alas! [W]e are only led on to inscrutable difficulties."[52] It was impossible to even imagine

> a finger-post pointing to a way that can lead us towards the explanation. That is not putting it too strongly. I only say we cannot now imagine it. But this time next year,—this time ten years,—this time one hundred years,—probably it will be just as easy as we think it is to understand that glass of water, which seems now so plain and simple. I cannot doubt but that these things which now seem to us so mysterious, will be no mysteries at all; that the scales will fall from our eyes. . . . [53]

With that attitude, it is difficult to believe that a letter of his in which he abandoned the vortex atom was the final word.[54]

Thomson's belief in the existence of the ultimate particle and the possibility of snaring it in a theory was not a vision of his. That belief was never made explicit in his earlier papers; he never expressed frustration or discontent with the mathematical analogies, models, or mathematical theories based on hypothetical physical concepts. It was only in his mature years, after 1884, that he looked back and recognized a pattern, a pathway to his thinking, leading to the goal he now identified as the description of the ultimate nature of matter and energy. That there was an internal consistency to his thought—a constant movement toward a real physical explanation—may seem uncanny since it was not conscious. But Thomson's success was due to his perseverance in following his own intuitions. The pattern of this thought, that is, the character of his mind, was gradually revealed. His research through all the areas of his many interests had an intellectual integrity.

When Thomson bewailed the suggestion that the ultimate explanation was not at hand, he was expressing the sentiment of a man who was emerging from a thicket and saw his goal within reach. He lived long enough to witness both the beginning and the end of a cycle of scientific theory. The dynamical theory, of which he was one of the founders, was reaching the limits of its explanatory powers. The apogee of the dynamical theory came near the start of the twentieth century at the moment when experimental evidence was showing its weaknesses as a unifying concept. It was a repeat of the scientists' construction of a conceptual Tower of Babel. When it almost seemed to touch the heavens, it was destroyed. Thomson's discomfiture was akin to the ancient builders of Babel.

CHAPTER THIRTEEN

A Natural Philosopher
And The New Physics

The latter part of Victoria's reign was a time of noticeable change in public attitudes that manifested itself in science. The pace of change was unsatisfactory to the scientific radicals who insisted that discoveries dealing with atomic structure of matter demanded drastic departures from the generalizations of the century and attention to particular facts. The new science emphasized the individual's work, and although no grand schemes resulted, each man's contribution stood by itself. Each scientist contributed discrete facts associated with but not dependent on each other. Röntgen discovered X rays, Becquerel found radioactivity, J. J. Thomson tracked down the electron, and the Curies isolated radium.

Home Rule, which excited the public mind in a way that was difficult to imagine in less particularistic times, was an issue between collective grandeur and individualistic expression. Were Irish interests separate from English? Could the Irish be satisfied in a state that expressed the common aspirations of British nationalism? The issue of Home Rule was of particular interest to Thomson, who was born in Ireland but lived most of his life in Scotland. He told an audience that he entered the agitation over this question in 1886, feeling compelled to become involved in political subjects that he found were far from his ordinary and natural avocations. In his opinion, Home Rule threatened to take Ireland away "from its grand position as a constituent equal member of the British Empire." Tyndall was also opposed to Gladstone and Home Rule.[1]

Thomson became president of the West of Scotland Liberal Unionist Association and was asked to help raise £20,000 for the central fund.[2] At one meeting of the General Executive of the Association he said:

> The condition of affairs would be intolerable for Ireland as it would be for England. It would be disastrous, ruinous for Ireland. (Hear, Hear.) It

216

would be most detrimental to the interests and Honour of England. Should this bill get so far as to pass through the House of Commons there would be an indelible disgrace upon the electorate of this country. (Cheers.)[3]

One of Thomson's nieces reported some of his political convictions. She wrote that Uncle William was for free trade and "it is evidently the Imperial idea which attracts him." At breakfast, he said that Benjamin Franklin was a staunch imperialist "ten years before the war of Secession" and that Britain lost the American colonies through mismanagement. The niece's diary entry also reported her version of Thomson's ideas on faction:

> Speaking of party politics Uncle William says party politics and party animosities do such frightful harm that he would be glad to see a break-up of the system of two parties in Parliament. He says, as things are at present, members have to vote for what they do not approve of for the sake of their party, and have to vote against what they want if it is proposed by the other side, and he thinks this most deplorable.[4]

In 1892, at the age of sixty-eight, Thomson was raised to the peerage. The name he chose was Kelvin, after the river Kelvin that flowed near the University of Glasgow. He was created Baron Kelvin of Largs. One journal commented, "A peerage has at last been conferred upon a scientific man." However, Kelvin was not only a "scientific man," for his wealth and fame came from his genius in combining science with practice.[5]

From the outset, Thomson's thinking, which was always on an abstract mathematical level, was tied to physical conceptualization. He had shown this inclination as a youth when he wrote his prize paper, "On the Figure of the Earth."[6] The requirements for the appointment to the chair of natural philosophy at Glasgow emphasized that he not be an "$x + y$ man,"[7] that is, that he be as knowledgeable in the laboratory as he was in theoretical mathematics. His trip to Paris in 1845 was a turning point in Kelvin's career. At that time, his apprenticeship in Regnault's laboratory not only acquainted him with laboratory technique but also directed his attention toward the practical application of theory.

Kelvin's involvement in the Atlantic cable scheme in his early thirties set the character of his long career. The whole question of the Atlantic cable was a test of scientific theory. Kelvin's interpretation of the theory of electric delay in long lines was verified. His invention of the mirror galvanometer overcame the disability that the delay caused. The success of his tide predictor confirmed Kelvin's views about the dynamics of gravity's effect on the earth. The mariner's compass was an invention that resulted from a discussion about magnetic lines of force. The devices that Kelvin designed were implementations of his scientific theories, and as in the case of the Atlantic cable, he was able to overcome the limitations of technology.

Kelvin's involvement in the Niagara Falls power project in the 1890s was in marked contrast to his role in the Atlantic cable. He had been approached in 1887 by S. P. Thompson to participate in a scheme to harness Niagara, but turned him down.[8] In 1890, however, Kelvin accepted the post of chairman of the International Niagara Commission, which judged an international competition for designing a power system for the falls. Kelvin's contribution to the Niagara project was minimal because the problem did not involve new scientific ideas—it involved two competing technologies. Scientific theory had no answer to the questions raised. The ultimate question at Niagara was one that centered on whether a.c. or d.c. would be used in transmitting power to Buffalo. To the very end, Kelvin insisted that d.c. was the only practicable way. He believed that the alternating current system had not been in use long enough, that it had not yet developed to a point of dependability.[9]

When Edward D. Adams, president of the company formed to handle the Niagara Falls power project, received Kelvin's cabled message, "Trust you avoid [the] gigantic mistake of adopting alternate currents," he undoubtedly thought, "Stubborn old man." S. P. Thompson put it another way. He wrote that Kelvin had "a curious impenetrability to other men's ways of regarding things. . . ."[10]

Kelvin's scientific stature grew at a compound-interest rate. His reputation increased as past work became acknowledged. Many of his early original insights became the foundation for the important scientific accomplishments in thermodynamics and electricity. Appreciation for Kelvin's originality rose as scientists recognized how basic his thought had been to the advancement of nineteenth-century science. Those who understood the importance of the first insight to an advance in thinking had reason to have more regard for Kelvin as time passed.

But Kelvin's modesty made him no more receptive to adulation than his reticence permitted him to enjoy using the weight of his reputation in controversy. That impulse to battle was exposed briefly at the age of seventeen when he attacked an established scholar for maligning Fourier. Dr. Thomson quickly stepped in and taught his son to couch his objections very carefully. Afterwards, Kelvin too often drew back when his opinion was most needed, as in the case of the Whitehouse crisis on the Atlantic cable, or he too willingly allowed the eager Tait to take up a quarrel on his behalf.

So it was that in 1894, when Kelvin's former student John Perry wanted to reopen the debate of the age of the earth, Perry had cause to complain, "my great difficulty is getting you to re-consider the question." Although Kelvin was silent at first, Tait's tactic was to try to brush Perry off. The sensitive Perry was certain that he was being ignored because he was not considered a worthy foe.[11] But Perry was wrong. Kelvin was incapable of hauteur. He was drawn into a controversy not by the stakes involved, but by the intrinsic interest of the question. As with former issues he raised, they became widespread debates about various specialty fields in the scientific community. Perry and others assumed that someone of Kelvin's stature would naturally want to test his strength in the contest. But Perry's efforts to rouse the giant were met by a mild reply.[12] Kelvin's correspondence with Stokes was another matter. There they dealt with intrinsically fascinating questions and were able to have a clash of opinion in the sheltered world of private correspondence.

Early in 1896, Stokes challenged Kelvin in his use of "normal vibrations" in his paper titled "On the Generation of Longitudinal Waves in Ether."[13] Besides, he thought that an experiment Kelvin had proposed would not be as conclusive as he had claimed. The outcome of the experiment depended on how electrostatic force affected the ether. Stokes did not know what this relation was, "and I am not aware that anyone else does; but I am not strong in electricity."[14]

In reply, Kelvin sent a formula cut from proof sheets of his Baltimore lectures. They had been in type since 1885, and because of Stokes's letter, Kelvin sent them to the Pitt Press. The formula was based on a model of a molecule surrounded by ether, and it described wave motion in the ether. This expression was Kelvin's idea of how the connection between matter and the ether was obtained. The model was "my realisation of the vibratile molecule which I first learned from you." He wrote to Stokes that he had "been trying almost incessantly since some time before Nov. 28, 1846 [see "On a Mechanical Representation of Electric, Magnetic, and Galvanic Forces," art. 27, vol. 1 of Kelvin's *Mathematical and Physical Papers*] for a mechanical-*dynamical* representation of electrostatic force, but hitherto quite in vain. Many others, including Maxwell, have tried more or less for the same, but with no approach to a glimpse of what might become a success." The letter was followed by a telegram. Kelvin to Stokes: "Formula yesterdays [*sic*] letter requires largely simplifying amendment not affecting our conclusions. I wrote by post. Kelvin."[15]

"On the Generation of Longitudinal Waves in Ether," first published in the Royal Society's *Proceedings* in 1896, was prompted by Wilhelm Röntgen's announcement a few months earlier of his discovery of X rays. The rays penetrated human flesh and produced a picture of the bones of a subject's arm on a photographic plate. Röntgen's first explanation of the X rays was that they were longitudinal waves similar to the propagation of sound waves. Kelvin wanted to test Röntgen's theory because it implied that the ether was compressible.[16]

The experiment that Kelvin proposed was a hypothetical one, that is, he had not performed it and did not know how to produce the effect he was looking for. He wrote: "It is not easy to see how this question could be answered experimentally; but remembering the wonderful ingenuity shown by Hertz," he hoped that someone would succeed with his suggestion. When performed, the experiment would show whether waves such as those made by X rays in the ether were longitudinal, proving that the ether was compressible, or nonlongitudinal, in which case the ether would have to be incompressible. The idea of an elastic solid ether, whether compressible or incompressible, was giving Kelvin trouble. He saw how such a substance served as a medium for light and other rays, but how were attractions as in the cases of electricity and magnetism produced by such a medium? He insisted that there must be an elastic solid "or a definite mechanical analogue of it, for the undulatory theory of light and of magnetic waves and of electric waves."[17] He believed that X rays had to fit into a dynamical, unified scheme of matter and energy.

Stokes did not like Kelvin's interpretation. A compressible ether implied an ether made of molecules. If so, how did action take place? In one ether, or were there two ethers? All of this reminded Stokes of the rhyme:

... and dogs have fleas that bite 'em,
And big fleas have little fleas, and so ad infinitum.

Stokes did not want to hypothesize about the structure of the ether: "My mind inclines me to scratch off the fleas, and rest in the idea of a plenum of incompressible ether, which our present knowledge does not authorise us in going beyond." Kelvin had cited an experiment by J. B. Perrin that examined interference produced by Röntgen rays. Stokes showed how the experiment might be explained in another way, but that interpretation would have to wait until Perrin provided more information about his experiment.[18]

Kelvin replied that it was harder for him to imagine an incompressible ether than finite compressibility. He did not see how Stokes could say that compressibility assumed either action at a distance or a second ether to explain it. "Therefore," Kelvin declared, "it seems to me difficult to believe that *there are not* longitudinal waves in ether. What do you say to this?" They were in agreement that Röntgen rays were "more probably extreme ultraviolet rays of transverse vibration than anything else we can think of just now." Kelvin suggested a joint note, which "will be a tremendous feather in my cap," to the Royal Society to this effect.[19]

The rapid exchange of ideas gave rise to letters crossing in the mail. The same day Kelvin wrote the above letter, 9 March, Stokes wrote a letter to Kelvin calling his attention to a paper appearing in the 2 March 1896 issue of *Comptes Rendus*. The explanation given in the paper, Stokes thought, was additional evidence of the smallness of the length of Röntgen rays and of the validity of Stokes's view. Kelvin's answer to Stokes's letter of 9 March was written on 12 March. He thought Stokes's idea a good one for conducting an experiment on the reflection of Röntgen rays from the surface of mercury. On 12 March, Stokes posted his reply to Kelvin's 9 March letter. In a postscript Stokes said, "I find that in the attempt to convey one's ideas to the mind of another, one's own ideas grow and get clearer."[20] Both men were over seventy; after fifty years, their correspondence was still lively.

In a letter of 13 March 1896, Stokes claimed that the "coup de grace" was given to normal vibrations in the ether by the experiments of Henri Becquerel and S. P. Thompson. As for the experiments on the reflection of X rays by Röntgen, M. M. Imbert, Bertin-Sans, and Lord Blythswood, they might have explanations based on totally different suppositions, that is, either the X rays were actually reflected or they were absorbed and the reflecting substance emitted other X rays. The difficulty might be resolved by conducting an experiment on lines described by Stokes. The length of this letter was eight typewritten pages. Stokes, whose handwriting was very difficult to read, was now typing his letters out of consideration for his correspondents. Halfway through the letter, there is another date, 14 March, and a note, "I resume before breakfast." There followed a detailed description of the means of conducting the test experiment, and he conjectured as to what each of the possible results would demonstrate.[21] It was becoming increasingly difficult to interpret the experimental results that were flooding in.

On 16 March 1896, Stokes wrote that he thought Röntgen's own experiments disproved the idea of the reflection of X rays. He felt "a good deal of con—," and the letter broke off at this point to be resumed on 19 March after Stokes had his typewriter repaired. In the meantime, Stokes received some exposed plates from Kelvin that were "a smashing blow to my [Stokes's] 'con.' " Another letter from Kelvin explained that photographs taken by a scientific amateur and friend, Lord Blythswood, were valueless because Blythswood's assistant had not followed instructions. After this reversal, Stokes felt, "I may almost complete my word with 'fidence.' " Henri Becquerel also claimed to have established the reflection of X rays, but Stokes thought that his report in *Comptes Rendus* could be explained by the hypothesis of absorption and reflection.[22]

Kelvin's note to the Royal Society was appended to one in which both Stokes's and Kelvin's views were given. The effort to write a note representing both men's views refined their ideas. After reading Kelvin's draft, Stokes objected that, "What you said about my views does not exactly represent them." The revision he suggested was the version that appeared in the *Proceedings of the Royal Society*.[23]

Röntgen's experiments extended the possible interpretations of ether waves instead of sharpening understanding of the ether by limiting possibilities. Stokes admitted that he had "rather turned from the idea" of rapidly vibrating molecules setting up ether waves to the idea of each molecule producing a wave pulse. This latest opinion fitted Stokes's theory of the propagation of light in the ether. At the end of March 1896, an article in *Comptes Rendus* reported the polarization of Röntgen rays. These results meant "assuming the correctness of the experiments," Stokes wrote, and that the rays were transverse vibrations in the ether. He added, "I have always myself been in favour of transversal vibrations of very high frequency." After reading the article, Kelvin replied, "So I suppose we may consider it as settled [by the experiments] . . . that Röntgen rays consist of transverse vibrations!! I am glad we had both come to that conclusion as probable."[24]

But reflection still remained a point of difference. Stokes argued that Röntgen himself thought the rays were sent back from surfaces of metals by something other than reflection. Of the two possible explanations for returning rays, Stokes favored the idea of what he called X fluorescence rather than straight reflection. After Stokes saw some new photographs, he confessed, "there *is* regular reflection, albeit very small. . . . " That was enough of an admission for Kelvin who wrote, "I am glad to hear that you think it [the experiment] proves a small proportion of the light to be regularly reflected!"[25]

After the initial surge of interest, Röntgen rays gradually passed from the center of interest. Letters between Stokes and Kelvin slowed. On 24 April, Kelvin wrote asking, "Don't you think Galitzin and Karnojitsky are to be trusted as having *seen* an effect by crossed tourmalines, though two or three others have tried for it and *not seen* it?" 19 June: "In train for London," Kelvin wrote to Stokes that he had tried an experiment on "X-light" [Röntgen rays] that he had read about. A book was supposed to produce retardation of the light, but Kelvin "found *none*; none perceptible."[26] By October 1896, the two men had turned their attention to phenomena called ice halos.

The last decade of the nineteenth century produced a juncture of remarkable discoveries. In addition to Röntgen's discovery of X rays, there was the subatomic particle, the electron, as a cause of cathode-ray phenomena as well as the whole realm of radioactive substances. Many thought the age was on the verge of an immense breakthrough. In his *History of European Thought in the Nineteenth Century*, John Theodore Merz thought

> all these scattered fragments or glimpses of knowledge promise at the end of the century to come together into a consistent theory of the nature of electricity as an atomically-constituted substance which is associated with particles of ponderable matter, or may even be the ultimate constituent of such matter itself.[27]

Some scientists enjoyed only a brief euphoria. George Fitzgerald of Trinity College, Dublin, wrote to Kelvin in March 1896:

> At one time I had great hopes that these Röntgen Rays would turn out to be related to gravitation in the same sort of way that light is to electromagnetism but I fear you are right and that they are only very rapid ultra violet light.[28]

Maxwell, had he lived, would have seen the effort as a phase of man's perennial quest. Some of these difficult questions were: Is space infinite and if so in what sense? Is the material world infinite in extent and are all places within that extent equally full of matter? Do atoms exist, or is matter infinitely divisible? These were hard questions, Maxwell said, with which man has perplexed himself:

> The discussion of questions of this kind has been going on ever since men began to reason, and to each of us, as soon as we obtain the use of

our faculties, the same old questions arise as fresh as ever. They form as essential a part of the science of the nineteenth century of our era, as of that of the fifth century before it.[29]

When an attempt to correlate all the data indicated there was still as much confusion as enlightenment about ultimate matter, the usual reaction of scientists was, as in the past, that it was the job of philosophy to work out the grand scheme; scientists could only hypothesize about one or two phenomena at a time. Kelvin was among the very few who did not think the discoveries revolutionized science. He thought the mission of science was to piece together all the phenomena of nature in the belief that all were related and that the grand scheme existed. He accepted the general view that the data had thus far allowed them to make only limited statements about the physical world. In his scientific writing, he carefully limited his speculations (some scientists said not enough) to what the information allowed him to say. But there was no restraining his imagination or his belief that what was known for certain was a real part of what could be only guessed at. He said it was man's purpose to attempt to know how all phenomena were related and what the ultimate matter was, even though man could only come closer and closer to the goal without reaching it.

When the Baltimore lectures were finally published in 1904, a series of articles were included as appendices in the volume. In these papers, Kelvin contended with the difficulties for the dynamical theory that were emerging. Appendix A to the published lectures was the reprint of a paper he had read to the Royal Society of Edinburgh in 1900. It was a reversion to abstract mathematical dynamics once again, avoiding physical hypotheses in order to deal in the abstract with a difficult physical question about the motion of atoms through the ether.[30] Mathematically, the theory was shown to be valid, but Thomson had to admit that the Michelson-Morley experiment presented "one serious, perhaps insuperable, difficulty." The two "coefficients" at Baltimore, Michelson and Morley, had conducted a series of experiments in 1887 in which they measured the speed of light in different directions in order to try to detect the effect of the ether on the light. There was no effect, and the conclusion seemed inescapable that the ether did not exist *unless,* as Thomson suggested, the theory broached independently by Fitzgerald and Lorentz was true. Their theory was that Michelson and Morley had not detected a change in the velocity of light because their instruments were foreshortened due to physical contraction. But the idea of physical contraction with increased velocity was a relativistic concept that no one was prepared to accept in 1900. The change in thinking which heralded the new physics, the beginning of the new cycle, occurred sometime after 1910.[31]

"Nineteenth Century Clouds over the Dynamical Theory of Heat and Light," Appendix B of the Baltimore lectures, was the reprint of a paper that Kelvin gave at the Royal Institution in 1900. He admitted that one of the most threatening clouds was the difficulty of conceiving of the earth and other solids as moving through the ether. Desperate measures were in order to save the ether. He suggested that the "scholastic axiom" stating that two bodies cannot occupy the same space at the same time be denied! Then all that was needed was to devise some new properties for the ether in order to account for the wave motion of light. Kelvin confessed that the exercise stretched his credulity: "I am afraid we must still regard Cloud No. 1. as very dense."[32]

The ether was the keystone to Kelvin's unifying concept and was difficult to set. Did the dynamical theory require the ether to be imponderable or did it have a mass that obeyed the gravitational law? He had told the physicists in Baltimore that there was no reason to believe that the ether was imponderable. A note added in 1899 retracted that assertion—the ether was imponderable. At the 1901 meeting of the British Association at Glasgow, he reiterated his belief in an imponderable ether with a wry reference to comtemptuous youth:

> I remember the contempt and self-complacent compassion with which sixty years ago I myself—I am afraid—and most of the teachers of that time looked back upon the ideas of the elderly people who went before us, and who spoke of "the imponderables."

He said it was necessary to look to the past, the dark ages of one hundred years before, for the truth of the belief in imponderables.[33]

The object of the Baltimore lectures had been to "find how much of the phenomena of light can be explained without going beyond the elastic-solid theory." Kelvin now had the answer: *everything nonmagnetic and nothing magnetic.* The electromagnetic theory of light had been of no help in finding a general theory of the ether, but Kelvin declared, "The grand object is fully before us of finding a comprehensive dynamics of ether, electricity, and ponderable matter, which shall include electrostatic force, magnetostatic force, electromagnetism, electrochemistry, and the wave theory of light."[34]

Twenty years went by as the manuscript of the Baltimore lectures passed through the printers, and the delay was Kelvin's. He persisted in thinking that the last piece of the puzzle was at hand, but it eluded him. After all this work, the volume was not a final statement. There could be no such thing for a man who may have stuck by his mode of thinking about physical phenomena but who constantly modified his theories. At the end of the twentieth lecture which he had rewritten, Kelvin came to some interesting questions of energy that would, when "fully worked out, include a dynamical theory of phosphorescence." But there was no time. "For the present I must leave it with much regret, to allow this Volume to be prepared for publication."[35]

Kelvin's fiftieth jubilee as professor of natural philosophy at Glasgow occurred in 1896. Was there more surprise than satisfaction among his friends when it was found in 1896 that Kelvin, the most restless of scientific intellects and world traveler, had occupied the same academic chair for fifty years? The Queen and the Prince of Wales sent greetings to the affair that was treated with pomp at Glasgow and lasted for three days. What made the occasion unique was that no other man could have deserved greetings from so wide a scientific and engineering circle. Messages were read from all the major British engineering and scientific societies that were pleased to mention their close ties with Kelvin. Universities in Britain and abroad sent greetings as did the leading men in science and engineering. Kelvin was particularly pleased with greetings he received from some former students in Japan and from the Baltimore "coefficients." In response to a toast Kelvin said:

> One word characterises the most strenuous of the efforts for the advancement of science that I have made perseveringly during fifty-five years; that word is FAILURE. I know no more of electric and magnetic force, or of the relation between ether, electricity, and ponderable matter, or of chemical affinity, than I knew and tried to teach to my students of natural philosophy fifty years ago in my first session as Professor. Something of sadness must come of failure; but in the pursuit of science, inborn necessity to make the effort brings with it much of the *certaminis gaudia.* . . . [36]

Many have been struck by the word that Kelvin stressed: FAILURE. At the time of the jubilee, he had been plunged again by the furor over Röntgen's discovery of considering the fundamentals of his approach to scientific theory. His involvement with these problems prevented him from seeing that his audience might misinterpret his remarks. More revealing were his words: "inborn necessity to make the effort." When he wrote these words, he was no doubt thinking of his father and J. P. Nichol. From his father, he learned the meaning of perseverance and dedication. From Nichol, the influence of his "creative imagination."[37] In a talk Kelvin gave in 1885, their influence is shown:

> Every person who aims at one object of course perseveres until he attains it; but he keeps his mind open until he can return to some other object never thought of at first, but which thrusts itself on him as a difficulty occurring in the pursuit of the first object. The very disappointments in attaining objects sought after in the investigations of physical science are the richest sources of ultimate profit, and present satisfaction and pleasure, notwithstanding the difficulties and disappointments contended with.[38]

Kelvin sent a set of proofs of the Baltimore lectures to Fitzgerald. In his answer, Fitzgerald reminded him that his "immense authority" gave him the responsibility to be careful about the way he presented his ideas; otherwise, wrong ideas about the elastic solid theory and electric force would be perpetuated.[39] Kelvin's correspondence with the younger physicists, such as Fitzgerald and Heaviside, bore little resemblance to the Kelvin-Stokes correspondence.

That correspondence began in 1846. Kelvin and Stokes were contemporaries who shared presuppositions. The same was true of his correspondence with Tait. However, neither Stokes nor Tait could have had the interchange with the young physicists that Kelvin had. As secretary of the Royal Society for thirty-one years, Stokes was an exemplar. But he was too cautious a man to expose himself in correspondence with younger men.[40] As for Tait, it is doubtful that many of the younger physicists would have exposed themselves to his wit and ready barb. Kelvin was a stimulating participant at scientific gatherings. He never hesitated to say what he thought. Though he may have infuriated the younger men, they wanted to engage him in a discussion or correspondence because he met them on equal ground. When Kelvin used the words *certaminis gaudia* ("the joys of the struggle") in his jubilee address, he meant exactly that. Otherwise, he would not have continued his correspondence with scientists who were of a different era.

Fitzgerald challenged Kelvin, who agreed to some changes in the Baltimore lectures, but he drew the line beyond which he would not go; Kelvin wrote: "It is mere nihilism, having no part or lot in Natural Philosophy, to be contented with two formulas for energy, electromagnetic and electrostatic, and to be happy with a vector and delighted with a page of symmetrical formulas."[41] All mathematical analysis must be founded on physical phenomena, *and* the real object of scientific investigation was to find the underlying physical explanation for all phenomena.

Fitzgerald agreed that one should not be satisfied with one or two equations as a description of phenomena. At the same time, he thought it "still further removed from Philosophy to say 'we must' when we need not." He was referring to Kelvin's statement in the Baltimore lectures that read: "The luminiferous ether we must imagine to be a substance which so far as luminiferous vibrations are concerned moves as if it were an elastic solid."[42] To which Fitzgerald replied: "Now this 'we must' is entirely unjustifiable. We *need* do nothing of the kind. There are lots of other possible changes in a medium which could be *represented*. . . . " He reminded Kelvin of how tentative he had been about some ideas and that he had occasionally inserted "probable" in his descriptions. Fitzgerald admitted that he had no idea that electric and magnetic forces operate in an ether, and he was "neither satisfied nor happy nor contented" with any explanation given so far. Fitzgerald knew the ether was not an elastic solid, but it had the required properties to transmit the electric and magnetic forces. The ether was acceptable to Fitzgerald, but, he wrote: "I cannot see, nor apparently can you, how it can be both electric and magnetic and at the same time an elastic solid."[43]

The exchange of letters between Fitzgerald and Kelvin during that spring of 1896 was a debate between men who could not agree. Kelvin retreated briefly to the position of "older man"; he advised Fitzgerald to "read, mark, learn, and inwardly digest Maxwell. . . . I think that when you have done so you will see that the statements marked || on the proof are correct, and will reconsider the whole of your letter to me. . . . " He received the following unabashed reply from Fitzgerald: "I don't agree at all with you. I don't think the statement you marked || is at all true."[44]

Fitzgerald pointed to the fundamental difference between them. He believed that all the phenomena involving the ether could be explained "by things to which we can attribute any convenient properties we please," while Kelvin had "a magnificent faith in the simplest doctrine." Fitzgerald could not arrive at any such faith: "I am afraid the relation of matter to ether by which electric and magnetic forces on matter are produced are a very much more complicated question than the discovery of a workable structure for the ether. They involve a workable structure for matter as well and I am afraid we are a very long way indeed from that. Matter seems to be so horribly complicated." With so many possibilities, Fitzgerald "objected to saying 'must' about the question at all."[45]

In 1898, the debate continued and had sharpened to a single point of difference. Fitzgerald wrote that as far as he was concerned, the elastic solid analogy for ether was "a *mere* analogy." "In our present state of ignorance," he added, "I much prefer to acknowledge that the elastic solid is a very imperfect analogue to the ether, than to complicate the analogy with an addendum for which there seems to me no corresponding phenomenon."[46] In answer, Kelvin sent a short note:

28th Nov. [18]98[47]

Dear Fitzgerald

It is certainly not an allegory on the banks of the Nile. It is more like an alligator. It certainly will swallow up all ideas for the undulatory theory of light, and dynamical theory of E & M. [electricity and magnetism] not founded on force and inertia. I shall write more when I hear how you like this.

Meantime I am off to London till Wednesday night, to return for my lecture on Thursday morning.

Yours,

Kelvin.

"I am not afraid of your alligator," replied Fitzgerald, "which swallows up theories not founded on force and inertia." Although he claimed he could be convinced, for the time, he would continue "[his] sitting on the fence attitude." Like so many others of the young physicists, Fitzgerald was waiting for "*some* experimental evidence . . . before [he] complicate[d] [his] conceptions therewith" [that is, with the idea of an elastic solid].[48]

The differences Kelvin had with Fitzgerald were very much like the difficulties he had with Maxwell's theories. To Kelvin, science was the search for general unifying principles. All concepts must eventually conform to this universal, whereas men such as Maxwell and his successors—Fitzgerald among them—were more positivistic in their approach. Maxwell had said there might and ought to be more than one way of looking at particular phenomena. Fitzgerald spoke of using "any convenient properties we please" and complained that there was too much to know about the nature of matter itself for anyone to venture generalizations. To Kelvin, this flexibility was heresy. The recent discoveries of radioactivity confirmed the positivistically inclined scientists in their conviction that generalization was foolhardy because it was based on too little evidence and would, no doubt, be upset by the next experimental report.

Kelvin's unwavering belief that nature must be thought of as a whole and that each investigation had a place in a larger scheme marked him as more of a philosopher, or rather a natural philosopher. This approach, Maxwell would have said, was perfectly acceptable as one of many possible views. And Maxwell did accept all that Kelvin said as both possible and provocative, but Kelvin was unable to appreciate fully Maxwell's ideas. The new generation of scientists, Maxwell's successors, were less flexible, so that contention resulted when each side rejected the other's views altogether.

Another member of the young group of physicists whom Kelvin regarded as a nihilist was Oliver Heaviside. The young physicists would probably want to have been known in the twentieth century as positivists. Heaviside had an unusual intellect, and a combative one. His mathematical interpretation of Maxwell's work was published in one of the technical journals of the day, and then publication of subsequent articles was suspended. The editor said his mathematics was incomprehensible to the readers; Heaviside maintained that the articles were suspended for personal reasons. Kelvin made special mention of Heaviside's work in his inaugural address as president of the Institution of Electrical Engineers in 1889. He also nominated Heaviside to the Royal Society.

The two men had occasion to correspond in 1904, when Kelvin sent Heaviside a copy of his paper "On Deep-Water Two-Dimensional Waves produced by any given Initiating Disturbance." The mathematics, according to Heaviside, was very complicated, and he sent some suggestions for simplification. There were many questions; at one point Heaviside wrote: "I don't understand that physically at all." The letter was peppered by "I cannot find," "I do not see why," and "It seems very extraordinary." One phrase touched off a sharp response from Kelvin. For one set of conditions Heaviside remarked: "And so far as I can see, there is no solution possible. . . . " "Don't say *that*," Kelvin exclaimed. "You and I both know that there *is* a solution though we may not see how to find it. *It is not possible that there could be no solution.*"[49] For the new school, of which Fitzgerald and Heaviside were forerunners, that no solution was possible was the only honest statement in the face of the unsettling information being supplied by the experimentalists such as Röntgen, Becquerel, and J. J. Thomson.

The sharp difference of view did not end the mathematical discussion that was so interesting to both men. As for Kelvin's underscored statement, Heaviside replied, it did "not allow for any special meaning. There is no solution in the sense I described, as is shown by all the formulae I have examined. . . . "[50]

Nothing more was said about the matter. In his reply, Kelvin concentrated on the "very interesting" mathematical problems. Like Fitzgerald and others of the new generation of physicists, Heaviside found Kelvin to be a good starting point. There were many ideas and questions in his work to take up in new research. They need not and would not accept his view that a general theory of all physical phenomena was possible and indeed necessary. Kelvin thought from the first that the dynamical approach was the key to this general statement. At the turn of the century, the physicists thought that at the state of "ignorance" no grand scheme involving light, electricity, magnetism, and the ether was possible. Kelvin would have gone so far that it was not possible *for the moment*. In spite of this fundamental difference of view, the young physicists found that what radiated from Kelvin's central theme was exceedingly suggestive.

The attempt at the turn of the century to draw closer to a general theory involved Kelvin in a controversy. The man who shunned controversy found himself involved in it on all sides, and he apparently was not disturbed. There was even a pointed disagreement with Stokes.

A debate began in November 1898 when Stokes wrote to Kelvin "about a point which I think we are not *quite* agreed." The point was the question of motion of an incompressible fluid when the boundary presents a sharp salient angle. Stokes disagreed with Kelvin's position that the angle "must" be regarded as absolutely sharp only as a limit to reduction of the bluntness indefinitely. "I say," Stokes wrote, "we *may* regard an infinitely sharp angle as the limit of blunted angle as the bluntness decreases indefinitely; but we are not *obliged* so to regard it." A letter from Stokes at the end of December noted that they agreed on most of the theory, but "on one point we distinctly differ." Stokes thought that Kelvin "always shirked the perfectly sharp edge, or rather refused to accept one of my postulates. . . . " And "I hold that the IDEAL problem I posed is perfectly rational, and that it has a solution." The next day Stokes noted: "The difference between us lies in a nutshell, but the shell is too strong for my crackers." Was the motion in a minimum kinetic energy solution (called MIKE by Stokes) stable, as Kelvin thought, or not, as Stokes thought? "I believe," Kelvin replied, that "the radius of the 'nutshell' containing the difference that will remain between us will prove to be as small as the radius of curvature of any edge however sharp that we may agree to think of." [At the bottom of the page he wrote: "And I don't care if the shell is then uncrackable!!"][51] A diagram shows that their discussion was an idealized version of the problem of objects moving through the ether. What Kelvin was trying to find was a mathematical answer to the problem of particles of matter moving through the ether.

The exchange continued through the following months with some opportunity to discuss the question face to face. The nutshell of difference remained. The following is one of the shortest letters in the exchange:

Lensfield Cottage, Cambridge, 7 Jan. 1899.[52]

My dear Kelvin,

A line to say that I retract what I said about discontinuity *coming on* in the absence of a sharp edge, as I am not clear about it. Supposing however it already exists (say in the case of a sphere) does a surface of discontinuity continue, being continually paid out, or is it washed away? I know what your answer would be, but query? are you right?

It is an exceedingly difficult matter (at least so it seems to me) to prove that the mike [minimum kinetic energy] motion is stable or unstable, whichever be the truth.

Yours sincerely,

G. G. Stokes

They attacked the question so thoroughly that finally they were able to see clearly what each himself believed, and they were forced to be explicit in stating their presuppositions. As Stokes said, they arrived at a point where "I doubt if either of us could give a mathematical demonstration of the correctness of the opinion which he holds." In his letter of January 1901, Stokes wrote that he had decided "perhaps after all there is not much, if any, difference between us in the matter of finite slip."[53]

In 1899, after fifty-three years of a fruitful and happy association with the University of Glasgow, Kelvin felt compelled to resign. The chair of natural philosophy had been ideally suited to his scientific temperament. The long summer break gave him time to travel and to work in his mathematical notebooks. The natural philosophy laboratory had led to his close association with White. Their collaboration began with the making of laboratory instruments and evolved into a partnership in the manufacturing of equipment. Kelvin gave as a reason for refusing all blandishments to leave the University of Glasgow its ideal and convenient setting for doing both his mathematical and practical work.

He had begun his long tenure on 1 November 1846. His letter of resignation to the principal gave as the immediate cause the reorganization resulting from the appointment of Dr. Magnus Maclean to the Technical College. Kelvin thought it would be best for future planning if a younger man occupied the chair.[54] However, he remained as a "research student." He and Lady Kelvin moved out of their university residence, and their furnishings were moved to Netherhall.

The restless and enthusiastic lecturer would no longer hear the cheering students, as he had first heard them at the end of his first year in the chair. Kelvin wrote to Heaviside that it was "a very troubled time." Early in 1900, he and Lady Kelvin bought a house in London.[55] Now, instead of lecturing from November to May, that time would be spent in London. His house would be a convenient base for meetings, lectures, and other events—activities to keep a restless man busy.

In June 1901, the University of Glasgow celebrated its ninth jubilee. Kelvin wrote to Tait that the jubilee committee had asked him to speak on "a very trite subject": James Watt. Ordinarily, this should have brought a sharp retort from Tait, but his son had been killed in the Boer War, and he was broken in spirit and health. In July, Peter Guthrie Tait died. Their friendship and writing of the *Treatise on Natural Philosophy* had begun together. "The two northern wizards," as Maxwell called them,[56] created a work that permanently tied their names together. The treatise was not written without pain. "I cannot say that our meetings were never unruffled," Kelvin wrote in his obituary of Tait. "We had keen differences. . . . We never agreed to differ, always fought it out. But it was almost as great a pleasure to fight with Tait as to agree with him. His death is a loss to me which cannot, as long as I live, be replaced."[57]

Earlier that year, Kelvin endured the loss of another worthy opponent, albeit a much younger one: George Fitzgerald. His death at the age of forty-nine was a shock to all.[58]

In Kelvin's life, 1902 was a brighter year. He and Lady Kelvin came to the United States. The ostensible reason for the trip was to visit the Eastman works in Rochester, New York; he was a vice-chairman of the Kodak Company of London. It was their fourth trip to America. (They made their third trip in 1897, when the British Association met in Toronto.) On each of the trips, Kelvin visited Niagara Falls because the potential of the falls fascinated him. He told a group in New York that sometime in the future Niagara's power would be harnessed to supply the power needs on the whole continent.[59]

When their ship arrived in New York on 19 April 1902, reporters were waiting to interview Lord Kelvin. He gave his views on wireless telegraphy—"one of the world's most remarkable inventions"—and air ships—"They never will be able to use dirigible balloons as a means of conveying passengers from place to place." The reporter observed that Kelvin was deaf and feeble. His hearing was indeed poor, and while he did look feeble, his actions belied his appearance. On the same day of his arrival in New York, he participated in the installation of the incoming president of Columbia University, Nicholas Murray Butler. When Kelvin was spotted in the procession by a student, the shout went up, "Hats off to Kelvin."[60]

Two days after his arrival, Kelvin was feted by more than 2,000 people at a reception sponsored by the American Institute of Electrical Engineers, Columbia University, the American Association for the Advancement of Science, the American Physical Society, the New York Academy of Sciences, the American Mathematical Society, and the Astronomical and Astrophysical Society of America. The affair was held in the Columbia University gymnasium. *Nature* reported to its readers that the Americans had been enthusiastic in their reception of Lord Kelvin. *Harper's Weekly* published a full-page picture of him and described the highlights of his life for its readers.[61]

The New York Edison Company sponsored a dinner at Delmonico's in his honor. Kelvin's picture was on the menu cover surrounded by pictures of his electrical apparatuses. A picture of the *Great Eastern* was also included. Little electric lights, hidden in the flowers on the table, blinked on and off. The guests represented various fields of electrical technology. The evening celebrated Lord Kelvin, the engineer. As the New York *Sun* wrote, "There are few instruments used on land or sea that do not owe something or everything to Lord Kelvin's active brain."[62]

When Kelvin testified before a committee of the House of Representatives, it was looked upon as an occasion because he was a member of the British nobility. For him, however, the occasion presented an opportunity to speak on behalf of something that he had long sought to have done in Britain. Twenty years before, Stokes had chided him for his commitment to ousting the "good old English inch." Kelvin testified before the House Committee on Coinage, Weights and Measures in support of a bill to make the metric system the standard of weights and measures in the United States. He approved of the committee's plan for a period of adjustment to the new system.[63]

His trip to America was marred by the pain he suffered from

facial neuralgia.[64] His first bout with it began about 1897 and had eventually become chronic. Its cause was unknown. Bouts became increasingly painful and sometimes incapacitating. But he could smile through his pain:

> Dictated to the dest of   16th Jan. 1905
> ny tressent steaking tower.
>
>   Dear Chrystal, [George Chrystal, general secretary of Royal Society of Edinburgh]
>
>   I an sorry to say it will not de tossidle for ne to de tressent at the R.S. to read ny tater this day week. Ny No. 5 denon [i.e., his fifth facial nerve] is still *very* aggressive. There are several letters of the althadet (c, t, p, and others) which I cannot tronounce without acute pain. And till I an nuch detter than I an now, steaking or reading a tater would de unintelligible, and would shock the hearers. . . . [65]

Early in 1903 Stokes died. "The world is poorer through his death, and we who knew him feel the sorrow of bereavement," Kelvin wrote. He recalled that Stokes taught him "the principles of solar and steller chemistry when we were walking about among the colleges some time prior to 1852." Shortly before, Kelvin had to vacate his fellowship because of marriage.[66] In recent years, he and Stokes had puzzled over the meaning of the various birth pangs of the new physics. Now, Kelvin was scientifically alone. He had lived long enough to know the pangs of being a survivor. Each death—Maxwell, Tait, Fitzgerald, and now Stokes—was a loss for what Kelvin cared about most, the world of scientific thought.

Kelvin viewed the work in radioactivity as a spectator and critic. Experiment for Kelvin had always been a way of testing theory that he had arrived at mathematically. Radioactivity was an "accidental" discovery in that it was not the result of an experiment that was testing a theory. After the discovery, a theory was searched for. Kelvin applauded the work of the experimentalists, although their approach was foreign to him. In 1903, he wrote to Henri Becquerel that he considered his work over the past six years "the most wonderful discovery which has ever been made in respect to properties of matter. . . . Nothing from the beginning of [the] world was previously known of properties of matter comparable in any way with your discovery of radioactivity, with all the consequences which, up to the present day, have flown from it."[67]

But in 1906, when Tait was no longer around to do battle, Kelvin wrote three letters to the *Times* opposing the "wild theories" about radium, specifically, the transmutation of elements, and Becquerel agreed with him that more decisive proofs were needed.[68] At a mathematical and physical conference of the 1906 British Association meeting, Frederick Soddy contended that by producing helium from radium he established that one element evolved from another. In a letter to the *Times*, Kelvin wrote that "an isolated experimental discovery by Sir William Ramsay and Professor Soddy brilliantly interesting as it is, and solidly instructive as it is towards the theory of radium, suggests nothing more towards any modification of the atomic doctrine proposed some 2500 years ago by Democritus and universally adapted by chemists and other philosophers in the 19th century. . . . "[69]

Oliver Lodge answered with a personal attack on Kelvin: "His brilliantly original mind has not always submitted patiently to the task of assimilating the work of others by the process of reading. . . ."[70] Kelvin replied: "I am quite sure my old friend Lodge could not wilfully be unjust to me, but I do not think he knows how carefully and appreciatively I have done all I could by reading, and by personal intercourse with many of the chief workers in the field, to learn experimental results and theoretical deductions regarding radio-activity, ever since its discovery ten years ago by Henri Becquerel. I scarcely think any other person has spent more hours in reading the first and second editions of Rutherford's 'Radio-activity' than I have."[71]

Tait would have said Kelvin was wasting his time. He would have considered the radioactivity debate as pointless as the discussion over Boscovichian atoms. Kelvin maintained stubbornly that if radium disintegrated, it simply did not fit the definition of an atom. He replied to one supporter in 1906 that

> the "disintegration of the radium atom" is wantonly nonsensical. It is nonsense very misleading and mystifying to the general public, because, if what is at present called radium can be broken into parts, it is not an atom. . . .
> I admire most sincerely and highly the energy of the workers in Radio-activity, and the splendid experimental results which they have already got by resourceful and inventive experimental skill and laborious devotion. I feel sure that as things are going on we shall rapidly learn more and more of the real truth about radium.[72]

Kelvin was seen as an obstructionist. J. J. Thomson complained that unlike Rayleigh and Stokes, who carefully considered a question before committing themselves, Kelvin was quick to give an opinion and invariably opposed new ideas. Rayleigh said that while Kelvin would generously praise experimental results, he was critical of theories. The young scientists were veering away from Kelvin to a new paradigm.[73]

Ernest Rutherford smarted under Kelvin's opposition, and that irritation was revealed in his account of a lecture he gave on radium at the Royal Institution in 1904:

> I came into the room, which was half dark, and presently spotted Lord Kelvin in the audience and realised that I was in for trouble at the last part of my speech dealing with the age of the earth, where my views conflicted with his. To my relief, Kelvin fell fast asleep, but as I came to the important point, I saw the old bird sit up, open an eye and cock a baleful glance at me! Then a sudden inspiration came, and I said Lord Kelvin had limited the age of the earth, *provided no new source was discovered.* That prophetic utterance refers to what we are now considering tonight, radium! Behold! the old boy beamed upon me.[74]

If Kelvin had been more discreet and mindful of his status, he might have remained silent during the turmoil when new theories were emerging. But at eighty-three, he was as unable to avoid the conflict of ideas as he had been when Faraday's and Joule's views were considered by most scientists contrary to accepted theory.

On 1 August 1907, Kelvin gave a paper titled "On the Motion of Ether produced by Collision of Atoms on Molecules containing or not containing Electrons."[75] It was still another attempt to bring together the many new experimental findings and the proposed theories into a concept about the nature of matter in its relation to energy. Kelvin's thoughts were not only daring and speculative, but also based firmly on many sources of information that he hoped to show were compatible. The many, often confusing and seemingly conflicting pieces of information, could be only related through metaphysics. The talk was that of a natural philosopher, and it was applauded by his listeners.[76] This was Kelvin's last public appearance.

In the fall, Lady Kelvin had a paralytic stroke. The distraught Kelvin, fearful that his second wife would also die and leave him wholly alone, hovered around the sick room. Lady Kelvin gradually recovered, but Kelvin, as a result of the strain, fell ill in November. He had what seemed to be a complete recovery, then fell ill again and gradually lost his strength. Lord Kelvin died on 17 December 1907.

The further a scientist's vision goes, the surer is his progress when dealing with immediate problems. The scientist who has a view of ultimate reality, whether that view be spoken or part of his private thoughts, is able to fit ideas about particular phenomena into a grand scheme. His theories are points along the path to a distant point. Interpreters of those theories, that is, the scientists who come later, sense the forward motion, although the resulting progress is often unexpected and many times in directions unacceptable to the originator. By holding to the early vision, the scientist finds himself being bypassed by younger scientists.

What happened to Albert Einstein fifty years after Kelvin's death was in many ways a repetition of that familiar story. Einstein refused to accept some of the consequences of quantum, just as Kelvin was disturbed by the implications of Maxwell's theory. However, Einstein's reputation was not as overshadowed as Kelvin's; in Einstein's case, his opponents recognized him as the originator of the movement of which quantum mechanics was a logical extension.[77]

Einstein's fierce debates with Niels Bohr were viewed by the scientific community with some amusement. But when Einstein persisted in a stubborn rejection of quantum mechanics as not fitting in with his grand scheme—"God does not play dice with the world"—surprise turned to derision. In 1942, Einstein wrote to a friend that "I am an old man mainly known as a crank who doesn't wear socks."[78]

In 1931, leading world scientists gathered to celebrate the centenary of James Clerk Maxwell's birth and to pay tribute to the man who had contributed an essential element to the change in conception of physical reality that was "the most fruitful that physics has experienced since the time of Newton." Those were Einstein's words.[79] The praise was well deserved.

Kelvin was hardly mentioned at the celebration, while Faraday received full recognition. The brief mention of Kelvin was made by Joseph Larmor, who by 1931 was an old man who had known Kelvin. It was Larmor who expressed appreciation for Kelvin and his ideas at the 1907 meeting of the British Association where Kelvin spoke for the last time. Larmor had said that Kelvin dealt with questions about the ultimate nature of electricity and the atom, but that Maxwell preferred to leave those difficult problems for a later time when others would solve them.[80]

The development of Maxwell's landmark electromagnetic theory of light owes much to the thinking of Faraday as well as Kelvin. Why is Faraday always mentioned in that connection while Kelvin is virtually ignored? Maxwell readily admitted his debt to Kelvin.[81] Ironically, Kelvin was being ignored because he did not accept Maxwell's theory. With the fading of Kelvin's image as the paragon of nineteenth-century science, the myth of Kelvin's recalcitrance toward new ideas was established.

The basis for this myth has come from S. P. Thompson's biography of Lord Kelvin. This book has been considered the definitive work on Kelvin's life. Almost without exception, every work that cites Kelvin relies on it. In the case of Kelvin and Maxwell's electromagnetic theory, what S. P. Thompson emphasized was Kelvin's rejection of the theory. But he did not wait for the publication of his biography to make his unequivocal view known. He stated it in his obituary of Lord Kelvin that was published in *Nature*.[82]

In his biography, Thompson was not satisfied to make his statement and go on; he repeated himself: "Sir William, as we know, had never accepted Clerk Maxwell's electromagnetic theory of light. . . . " "But Thomson . . . would not bring himself to accept Maxwell's views." "That he did not greet Clerk Maxwell's magnificent hypothesis with the enthusiasm with which it was acclaimed by the younger school of British physicists, is a fact not to be ignored or explained away."[83] But a few pages later, S. P. Thompson adds:

> Maxwell's own statement of the theory was not without difficulty. It rested on a hypothesis concerning the elastic medium, of which he himself, in his great book on *Electricity and Magnetism* of 1873, was not able to give a very satisfying account; and students of that work were greatly puzzled to attach any consistent physical meaning to the conception of electrical "displacements" which lay at the root of the matter. The fact is, that Maxwell's ideas also were in a state of flux, of which the abrupt transitions and gaps in his exposition remain to testify. Even now it is not possible to give a clear dynamical statement of that which he called displacement, except by reading into it the discoveries of recent years.[84]

Why did Thompson treat Kelvin's objection to Maxwell's theory in such a way? He himself answered that question: "Not in our generation will it be possible to exercise dispassionate vision or to disentangle the ultimate from the obvious."[85]

Logically, the estimate of a scientist's contribution to the advancement of science must take into account his life's work. There is a danger that this estimate will be prejudiced by the controversies of his old age. It would be a gross injustice and a historical distortion to see Einstein as "an old man mainly known as a crank."

One use of the history of science is to give a time sense to scientific ideas that exist outside of time. The life of William Thomson, Sir William Thomson, and Lord Kelvin stretched across an age when scientific thought was undergoing a revolution. To understand and appreciate the magnitude of that change requires that the origins and supercessions of those ideas be seen in the context of the period.

Was the new science different from the old? Was it "progress"? It was, to the participants who felt an exhilaration at the new things being discovered every day and who expected a continuation of the excitement. To the person with a historical perspective, the questions being raised about the ultimate nature of matter and the relation between matter and energy were reminiscent of the unanswered questions of the past. The ideas of the twentieth century were different because man's view of the world never could be the same following the generalizations of the nineteenth century. The hope that all theories would eventually come under the grand generalization of the dynamical view of nature was unrealized, but in that attempt, the scientists succeeded in formulating the great law of the conservation of energy, as well as the other law of thermodynamics. Maxwell's extension of Thomson's ideas on magnetism and electricity to the electromagnetic theory of light might stand alone as the most important accomplishment of the century. Lord Kelvin was sure that he knew the object of science and did not waver from the statement he made to his first class at Glasgow in 1846. The object of natural philosophy, he said, was the establishment of general laws by induction from the facts. Kelvin, despite his sense of frustration, had seen that hope realized as far as any general scheme of nature can ever be realized.

NOTES

Most of the manuscript material and letters used in this biography of William Thomson (W. T.) can be found in the Cambridge University Library (CUL); the collection is now called The Kelvin Collection (Add. MS 7342).

Unless otherwise noted, the letters and manuscript materials cited here can be found in The Kelvin Collection, CUL. This collection has been reorganized, and the manuscript materials and letters can be found, using the library's system.

Throughout the text, abbreviations within quotations have been written out by the author.

Preface

1. S. P. Thompson, *The Life of William Thomson: Baron Kelvin of Largs*, 2 vols. (London: Macmillan, 1910), 2:1086.

Chapter 1

1. Elizabeth T. King, ed., *Lord Kelvin's Early Home* (London: Macmillan, 1909), pp. 1 and 3.
2. Thompson, *Life*, 1:2–3; Kelvin, "Address on Installation as Chancellor of the University of Glasgow," in *Mathematical & Physical Papers*, 6 vols. (Cambridge: Cambridge University Press, 1882–1911), 6:371 (hereafter cited as *Math. and Phys. Papers*).
3. Alexander Morgan, *Scottish University Studies* (London: Oxford University Press, 1933), p. 78.
4. Thompson, *Life*, 1:3; Morgan, *Studies*, p. 78; King, *Kelvin's Home*, pp. 10–11.
5. King, *Kelvin's Home*, p. 50; Thompson, *Life*, 1:5–7.
6. King, *Kelvin's Home*, pp. 70–72.
7. Ibid., p. 90; Maria Edgeworth, *Harry and Lucy Concluded: Being the Last Part of Early Lessons*, 2d ed. corrected, 4 vols. (London: R. Hunter, 1827), 3:79–92. In her book, Elizabeth writes of how the Thomson children emulated *Harry and Lucy* and built a bridge. See King, *Kelvin's Home*, p. 86.
8. Ibid., p. 128.
9. Max Lerner, ed., *Essential Works of John Stuart Mill* (New York: Bantam, 1961), pp. 13–14, 15–16, 29.
10. King, *Kelvin's Home*, pp. 23, 103, and 118.
11. See "Biographical Sketch" in James Thomson, *Collected Papers in Physics and Engineering* (Cambridge: Cambridge University Press, 1912), pp. xiii–xci.
12. King, *Kelvin's Home*, pp. 91–92, 26–27.
13. Ibid., pp. 87–88.
14. Ibid., pp. 97–98.
15. Ibid., pp. 99–100.
16. Thompson, *Life*, 1:8–9.
17. King, *Kelvin's Home*, letter from W. T. to Elizabeth, 24 May 1836, inserted after p. 138.

18. Ibid, pp. 120–21.

19. Thompson, *Life*, 1:41n; King, *Kelvin's Home*, pp. 216 and 214.

20. Ibid, p. 43; W. T., 1839 Diary, 24 May 1839. An overall view of The Kelvin Collection and The Stokes Collection, CUL, can be seen in *Catalogue of the Manuscript Collections of Sir George Gabriel Stokes and Sir William Thomson, Baron Kelvin of Largs in Cambridge University Library*, compiled by David B. Wilson (Cambridge: Cambridge University Library, 1976).

21. King, *Kelvin's Home*, pp. 145–46, 149, 150–51.

22. W. T., 1839 Diary, 31 May and 5 June 1839.

23. King, *Kelvin's Home*, pp. 150, 155, 172, and 169.

24. Ibid, pp. 171 and 190; Emmerich Diary, 26 and 27 May 1840.

25. Lerner, *Mill*, p. 28.

26. Three of W. T.'s Glasgow notebooks are in The Kelvin Collection: "Natural Philosophy," "Moral Philosophy," and "Logic."

27. "Lord Kelvin and His First Teacher in Natural Philosophy," *Nature* 68 (October 1903): 624. This article is a report of a talk given by Kelvin, 17 October 1903, on the occasion of the unveiling of a stained-glass window at the University of Glasgow in memory of John Pringle Nichol. Kelvin is quoted extensively.

28. Henri Poincaré, *Science and Method*, trans. F. Maitland (London and New York: Nelson, 1914), pp. 59–60. See also Jacques Hadamard, *The Psychology of Invention in the Mathematical Field* (Princeton: Princeton University Press, 1945).

29. Thompson, *Life*, 1:13.

30. Lerner, *Mill*, pp. 45–47.

Chapter 2

1. Quoted from Nichol's obituary, *Annual Register* (1859), p. 465; *DNB*, 14:413.

2. King, *Kelvin's Home*, p. 145.

3. "Lord Kelvin and His First Teacher in Natural Philosophy," *Nature*, p. 624.

4. W. T., "An Essay on the Figure of the Earth," title page.

5. The works that Thomson consulted were: "Airy's Tracts, and his treatise on the Figure of the Earth, in the Encyclopaedia Metropolitana; the Chapters on the Equilibrium of Fluids, and Attraction, in Poisson's Traité de Mécanique; the Chapter on the Attractions of Spheroids in Pontécoulant's Théorie du Système du Monde; the Chapters on the Figure of the Earth, and the Lunar Theory in Pratt's Mechanical Philosophy, and the Mécanique Céleste of Laplace, Liv. VII, Chap. II," Preface to "An Essay on the Figure of the Earth."

6. Ibid, pp. 2–3.

7. Ibid, pp. 1–4.

8. Ibid, pp. 4–5, 57–70.

9. Ibid, pp. 70–85.

10. Ibid, facing p. 42, pp. 54–55, facing p. 49.

11. Thompson, *Life*, 1:14.

12. Examples of a disparity of views are: the controversy between Faraday and the French mathematicians over the explanation of electromagnetic force, as well as the conflicting ideas about the nature of heat expressed by Carnot and Joule. For a further discussion, see H. I. Sharlin, *The Convergent Century: The Unification of Science in the Nineteenth Century* (London and New York: Abelard-Schuman, 1966), pass.

13. W. T., *Math. & Phys. Papers*, 6 vols., 3:296; see also vols. 1–6.

14. Joseph Fourier, *The Analytical Theory of Heat*, trans. A. Freeman from the French original publication (New York: Dover, 1955), pp. 1–2, 20–21.

15. Ibid, p. 4.

16. For a survey of Fourier's life and work, see I. Grattan-Guinness in collaboration with J. R. Ravetz, *Joseph Fourier 1768–1830* (Cambridge, Mass. and London: M. I. T. Press, 1972).

Notes

17. Fourier, *Heat,* pp. 24–25.

18. Ibid., p. 16.

19. Thompson, *Life,* 1:17–18.

20. P. Q. R., "On Fourier's Expansions of Functions in Trigonometrical Series," *Cambridge Mathematical Journal* 2, no. 12 (May 1841):258–59 (hereafter cited as *Camb. Math. Jour.*).

21. W. T., original draft of the article "On Fourier's Expansions."

22. Gregory to Dr. Thomson, 28 February 1841; Kelland to Dr. Thomson, 1 March 1841; Kelland to Dr. Thomson, 8 March 1841; Dr. Thomson to Gregory, 9 March 1841.

23. P. Q. R., "Note on a Passage in Fourier's Heat," *Camb. Math. Jour.* 3, no. 13 (November 1841):25–27.

24. Ibid., pp. 25–26.

25. François Arago, *Biographies of Distinguished Scientific Men,* trans. W. H. Smyth, B. Powell, and R. Grant (London: Longmans, 1857), p. 269; Sharlin, *Convergent Century,* pp. 20–21.

26. Dr. Thomson to Gregory, 23 August 1841.

27. P. Q. R., "On the Uniform Motion of Heat in Homogeneous Solid Bodies, and its Connection with the Mathematical Theory of Electricity," *Camb. Math. Jour.* 3, no. 14 (February 1842):71–84.

28. W. T. to Gregory, n.d. [1841]; Dr. Thomson to W. T., 15 November 1841.

29. W. T., *Reprint of Papers on Electrostatics and Magnetism,* 2d ed. (London: Macmillan, 1884), pp. 1–2 (hereafter cited as *Elec. and Mag.*).

30. W. D. Niven, ed., *The Scientific Papers of James Clerk Maxwell,* 2 vols. bound as 1 (New York: Dover, 1965), 2:302.

31. Ibid., 2:301.

32. For a further discussion of the mathematical theory, see chapter 4; W. T., *Elec. and Mag.,* p. 26.

33. *Nature* 68 (October 1903): 624.

34. Thompson, *Life,* 1:19. Quotation: This was the way Kelvin described Fourier in *Nature,* p. 624.

Chapter 3

1. Green wrote his essay before he entered Cambridge.

2. Dr. Thomson to W. T., 28 October 1841.

3. Dr. Thomson to W. T., 12 January 1842; W. T. to Dr. Thomson, 15 January 1842, 30 March 1842.

4. W. T. to Dr. Thomson, 12 December 1841.

5. William Hopkins, *Remarks on the Mathematical Teaching of the University of Cambridge,* n.p., n.d. "Early in 1854" written on the title page of the Cambridge University copy, p. 27.

6. William Hopkins, *Remarks on Certain Proposed Regulations Respecting the Studies of the University and the Period of Conferring the Degree of B.A.* (Cambridge: J. & J. J. Deighton, London: J. W. Parker, 1841), pp. 12–14.

7. Hopkins, *Mathematical Teaching,* p. 42; Hopkins to Dr. Thomson, 18 January 1845; W. T. to Dr. Thomson, 12 December 1841; "Hopkins Mss.," 'Dynamics Newton,' p. 12.

8. W. T. to Dr. Thomson, 21 November 1841; n.d. [January 1842].

9. W. T., Cambridge 1843 Diary, Diary no. 1, 15 February 1843.

10. Dr. Thomson to W. T., 27 March 1842; Cambridge Diary, no. 1, p. 16, n.d., 23 February 1843; W. T. to Dr. Thomson, 5 May 1843.

11. William Whewell, *Of a Liberal Education in General; and with Particular Reference to the Leading Studies of the University of Cambridge* (London: J. W. Parker, 1845), pp. 44–45; W. T. to Dr. Thomson, 27 December 1841.

12. W. W. Rouse Ball, *A History of the Study of Mathematics at Cambridge* (Cambridge, 1889), pp. 120-21. See also J. M. Dubbey, "The Introduction of the Differential Notation to Great Britain," *Annals of Science* 19, no. 1 (March 1963): 37-48.
13. W. T. to Dr. Thomson, n.d. [February or March 1842].
14. *Camb. Math. Jour.* 1:1-2.
15. Ball, *Mathematics*, p. 130. See also Elaine Koppelman, "The Calculus of Operations and the Rise of Abstract Algebra," *Archive for the History of the Exact Sciences* 8, no. 3 (1971): 155-242; W. T. to Dr. Thomson, 5 December 1841.
16. W. T. to Dr. Thomson, 27 December 1841.
17. Dr. Thomson to W. T., 27 March 1842, 8 April 1842.
18. W. T., *Elec. and Mag.*, p. 126.
19. W. T., Diary no. 1, 16 and 17 March 1843.
20. W. T., 1843 Diary Continued, Diary no. 2, 1 May 1844. See also *Journal de Mathématiques Pures et Appliqués* 9 (1844): 239-44.
21. W. T., Diary no. 1, 5 March 1843.
22. Whewell, *Liberal Education*, pp. 44-45, 30-31, and 74.
23. Cambridge University Commission, *Report of Her Majesty's Commissioners appointed to inquire into the State, Discipline, Studies, and Revenues of the University and Colleges of Cambridge: Together with the Evidence and an Appendix* (London: Her Majesty's Stationery Office, 1852), p. 23; "Government Education Measures for Poor and Rich," *Edinburgh Review* 99 (January 1854): 186-87.
24. See H. I. Sharlin, *The Making of the Electrical Age: From the Telegraph to Automation* (London and New York: Abelard-Schuman, 1963), pp. 205, 207-10.
25. W. T., Diary no. 1, 24 March 1843.
26. Ibid., [15 March 1843], 14 March 1843.
27. Ibid., 23 and 24 March 1843; ibid., 13 and 17 April, 23 March 1843; Dr. Thomson to W. T., 23 Oct. 1842; W. T. to Dr. Thomson, 26 Oct. 1842; W. T. to Dr. Thomson, 19 February 1842.
28. Dr. Thomson to W. T., 21 February 1842; W. T. to Dr. Thomson, 25 February 1842; Elizabeth Thomson to W. T., 27 February 1842.
29. Arthur Gray, *Cambridge University: An Episodical History* (Boston: Houghton Mifflin, 1927), p. 275.
30. W. T. to Dr. Thomson, n.d. [February or March 1842]; W. T., Diary no. 1, 16 and 24 March, 16 and 24 April 1843.
31. Ibid., 23 October 1843.
32. L. Campbell and W. Garnet, *The Life of James Clerk Maxwell* (London: Macmillan, 1882), p. 154. All three articles appeared in the *Camb. Math. Jour.* and were signed P. Q. R.: "On the Attractions of Conducting and Non-Conducting Electrified Bodies" (May 1843); "Note on Orthogonal Isothermal Surfaces" (May 1843 and November 1844); and "On the Equations of the Motion of Heat Referred to Curvilinear Co-ordinates" (November 1843).
33. W. T., Diary no. 1 [15 March 1843], 2 April 1843.
34. Ibid., 20 and 27 April, 1 and 14 May 1843.
35. Ibid., [October and November 1843].

Chapter 4

1. W. T. to Agnes Gall, 11 December 1844.
2. 1841: "On Fourier's Expansions of Functions in Trigonometrical Series," "Note on a Passage in Fourier's *Heat*," 1842: "On the Uniform Motion of Heat in Homogeneous Solid Bodies, and its Connexion with the Mathematical Theory of Electricity," "On the Linear Motion of Heat," "Propositions in the Theory of Attraction," 1843: "On the Attractions of Conducting and Nonconducting Electrified Bodies," "Notes on Orthogonal Isothermal Surfaces," "On the Equations of the Motion of Heat referred to

Notes

Curvilinear Co-ordinates." 1844: "Elementary Demonstration of Dupin's Theorem," "Note on some Points in the Theory of Heat," "Note on the Law of Gravity at the Surface of a Revolving Homogeneous Fluid," "Note sur la théorie de l'attraction."

3. S. P. Thompson, *Life,* 1:98.

4. Ibid., p. 110.

5. W. Hopkins to Dr. Thomson, 18 January 1845.

6. Dr. Thomson to W. T., 20 April 1843. Forbes was professor of natural philosophy at the University of Edinburgh.

7. W. T. to Dr. Thomson, 17 January 1845.

8. King, *Kelvin's Home,* p. 230.

9. W. T. to Dr. Thomson, 18 January 1845; Dr. Thomson to W. T., 20 April 1843.

10. Dr. Thomson to W. T., 16 May 1846; W. T. to Dr. Thomson, 7 October 1844.

11. Dr. Thomson to W. T., 12 October 1844.

12. Ludwig Fischer to Dr. Thomson, 16 February 1845.

13. Dr. Thomson to W. T., 8 January 1845.

14. ———, "Paris in 1846," *Dublin University Magazine* 28 (August 1846): 179.

15. Ibid., pp. 183–84; see also Joseph Ben-David, "The Rise and Decline of France as a Scientific Centre," *Minerva* 8 (April 1970): 160–79.

16. Martin Malia, *Alexander Hertzen and the Birth of Russian Socialism* (New York: Grosset Universal Library, 1965), pp. 358–60; F. Fuller to W. T., 16 February 1845; W. T. to Dr. Thomson, 16 March 1845; Thompson, *Life,* 1:127.

17. W. T. to Dr. Thomson, 23 and 10 February 1845.

18. See H. I. Sharlin, *Convergent Century,* pp. 185–87.

19. Dr. Thomson to W. T., 12 February 1845.

20. Dr. Thomson to W. T., 4 and 12 February 1845.

21. Ibid.

22. Biographical information for Regnault in *Encyclopaedia Britannica,* 11th ed. (hereafter cited as *Ency. Brit.*); W. W. Randall, ed. and trans., *The Expansion of Gases by Heat: Memoirs by Dalton, Gay-Lussac, Regnault, and Chappuis* (New York: American Book, 1902).

23. W. T. to Robert Thomson, 5 March 1845; W. T. to Dr. Thomson, 16 March 1845; Dr. Thomson to W. T., 22 March 1845.

24. W. T. to Dr. Thomson, 30 March 1845.

25. W. T. to Dr. Thomson, 14 April 1845.

26. W. T. to Dr. Thomson, 10 February 1845.

27. W. T. to Dr. Thomson, 23 February 1845, 16 March 1845.

28. W. T. to Dr. Thomson, 10 February 1845.

29. See article on Cauchy in *Ency. Brit.,* 11th ed., and J. T. Merz, *A History of European Thought in the Nineteenth Century,* 4 vols. (New York: Dover, 1965), 2: 636–37.

30. W. T., 1843 Diary Continued, Diary no. 2, 4 February 1845; W. T. to Dr. Thomson, 10 February 1845.

31. W. T. to Dr. Thomson, 23 February 1845.

32. W. T., Diary no. 2, 4 February 1845.

33. N. M. Ferrers, ed., *Mathematical Papers of the Late George Green* (London, 1871), p. 4.

34. Ibid., pp. 6–7.

35. Ibid., pp. 4–6.

36. Ibid., pp. 7–8.

37. W. T. to Dr. Thomson, 30 March 1845.

38. Niven, *Scientific Papers,* 2:301–2.

39. W. T., "The Wave Theory of Light," in *Popular Lectures and Addresses,* 3 vols. (London and New York, Macmillan 1891: 1894), 1:322.

40. W. T. to Dr. Thomson, 30 March 1845.

41. W. T., Diary no. 2, 10 September 1844.

42. W. T., "I—On the Elementary Laws of Statical Electricity," in *Elec. and Mag.,* p. 26. Sir William Snow Harris (1791–1867) turned from medicine to the study of

electricity in 1824 and became a fellow of the Royal Society in 1831 on the basis of his electrical research. His 1834 paper "On Some Elementary Laws of Electricity" was his first major publication. In 1835 he received the Royal Society's Copley gold medal and was knighted in 1847. Although his work was considered important in his day, Harris's contributions to science did not have lasting historical importance.

43. Ibid., p. 29.

44. Ibid., p. 24. For additional information, see E. W. Whittaker, *A History of the Theories of Aether and Electricity*, 2 vols. (London: Thomas Nelson, 1951); J. J. Thomson, "Report on Electrical Theories," *Report of the British Association* (1885), pp. 97–155.

45. W. Snow Harris, "On Some Elementary Laws of Electricity," *Philosophical Transactions Royal Society* (1834), part 2, p. 245.

Chapter 5

1. W. T. to Dr. Thomson, 20 and 14 April 1845.

2. John Thomson to W. T., 26 April 1845.

3. Dr. Thomson to W. T., 22 March 1845; W. T. to Dr. Thomson, 16 and 30 March 1845.

4. John Thomson to W. T., 18 April 1845.

5. James Thomson to W. T., 10 May 1845.

6. J. Challis to W. T., 10 May 1845; W. T. to Dr. Thomson, 17 August 1845.

7. W. T. to Dr. Thomson, 28 June 1845; Thompson, *Life*, 1:134.

8. Dr. Thomson to W. T., 1 July 1845.

9. W. Bottomley to W. T., 5 July 1845; W. T. to Dr. Thomson, 29 June 1846.

10. James Thomson to W. T., 1 July 1845.

11. Robert Thomson to W. T., 8 July 1845; Elizabeth King to W. T., July 1845.

12. Thompson, *Life*, 1:144–46.

13. John Thomson to W. T., 2 July 1845.

14. Ellis to W. T., 13 June 1845; *Camb. Math. Jour.* 1:1.

15. W. T. to Dr. Thomson, 2 June 1844.

16. J. T. Merz, *A History of European Thought in the Nineteenth Century*, 4 vols. (New York: Dover, 1965), 2:673; 1:274–75.

17. W. T. to Agnes Gall, 28 July 1845; Ellis to W. T., 26 July 1845.

18. W. Bottomley to W. T., 5 July 1845; De Morgan to W. T., 12 November 1845.

19. Boole to W. T., 18 July 1845; Smith to W. T., 16 July 1845; Ellis to W. T., 24 July 1845; Grant to W. T., 1 August 1845.

20. W. T., 1845–1863 Diary, Diary no. 3, 5, 12, 14 July 1845, pp. 5–10. Diary no. 3 does not include January 1858 to 26 June 1859, which is Diary no. 4.

21. Ibid., 18 August 1845, pp. 11–12.

22. W. T. to Dr. Thomson, 17 August 1845; W. T., Diary no. 3, 9 September 1845, pp. 14–15.

23. W. T., "Electric Images," in *Elec. and Mag.*, pp. 154–77.

24. Thomson expanded on the idea in 1869 in "Determination of the Distribution of Electricity on a Circular Segment of Plane or Spherical Conducting Surface, Under Any Given Influence," Ibid., pp. 178–91; W. T., "On Electrical Images," *Report of the British Association* (1847), Transactions of the Sections, pp. 6–7. See Maxwell's appreciation in Niven, *Scientific Papers*, 2:302–4.

25. John Thomson to W. T., 9 October 1845; W. T. to Dr. Thomson, 5 October 1845.

26. W. T. to Dr. Thomson, 10 October 1845.

27. W. T. to Dr. Thomson, 19 October 1845.

28. Ibid.

29. W. T. to Dr. Thomson, 1 November 1845; W. T. to Dr. Thomson, 16 March 1845.

Notes

30. W. T. to Dr. Thomson, 11 February 1846; L. Wallich to W. T., 15 June 1846.
31. Thompson, *Life,* 1:202.
32. W. T., Diary no. 3, 1 July 1846, pp. 21–24.
33. Ibid., 1 July 1846, p. 23.
34. A. Smith to W. T., 11 December 1845; *Times* (London), 5 November 1845, p. 4.
35. Quoted in Maxwell's article on Faraday in *Ency. Brit.,* 11th ed.
36. *Athenaeum* (1846), no. 953, p. 126; W. T. to Dr. Thomson, 28 January 1846.
37. W. T. to James Thomson, 7 April 1846.
38. W. T. to Dr. Thomson, 10 May 1846; David Thomson to W. T., 25 October 1845.
39. Dr. Thomson to W. T., 8 February 1846; W. T. to Dr. Thomson, 11 February 1846.
40. Dr. Thomson to W. T., 26 March 1846; W. T. to Dr. Thomson, 28 March, 13 May 1846.
41. Airey to W. T., 20 September 1846; Thompson, *Life,* 1:74.
42. James Thomson to W. T., 23 May 1846.
43. See *DNB*; Boole to W. T., 17 August 1846.
44. Charles Babbage, *Reflections on the Decline of Science in England, and on Some of Its Causes* (London: B. Fellowes & J. Booth, 1830), pp. 36–37.
45. Fischer to W. T., 14 February 1847.
46. Dr. Thomson to W. T., 7 May 1846.
47. David Layton, *Science for the People: The origins of the school science curriculum in England* (New York: Science History Publications, 1973), pp. 49 and 19; W. T. to Dr. Thomson, 28 January 1846.
48. W. T. to Dr. Thomson, 10 May 1846.
49. Dr. Thomson to W. T., 13 June 1846; W. T. to Dr. Thomson, 17 June 1846.
50. Dr. Thomson to W. T., 2 May 1846.
51. Dr. Thomson to W. T., 7 May 1846.
52. Dr. Thomson to W. T., 26 May 1846; W. T. to Dr. Thomson, 21 May 1846; James Thomson to W. T., 23 May 1846; W. T. to Dr. Thomson, 10 June 1846.
53. W. T. to Dr. Thomson, 17 June 1846.
54. W. T. to Dr. Thomson, 27 June 1846.
55. James Thomson to W. T., 12 May 1846; W. T. to Dr. Thomson, 27 June 1846.
56. Dr. Thomson to W. T., 2 and 16 May 1846.
57. W. T. to Dr. Thomson, 17 June 1846.
58. W. T. to Dr. Thomson, 21 November 1841; A. Smith to W. T., 14 May 1846.
59. Dr. Thomson to W. T., 21 June 1846.
60. A. Smith to W. T., 14 May 1846; W. T. to Dr. Thomson, 21 May 1846; A. Smith to Dr. Thomson, 10 June 1846.
61. Dr. Thomson to W. T., 8 July 1846.
62. Agnes Gall to W. T., 27 February and 9 September 1846; W. Bottomley to W. T., 15 June 1846.
63. E. T. King, *Kelvin's Home,* pp. 232–33.
64. Alexander Morgan, *Scottish University Studies* (London: Oxford University Press, 1933), p. 4.
65. Ibid., p. 172.
66. ———, *Report of the Royal Commissioners Appointed to Inquire into the Universities of Scotland with Evidence and Appendix.* Presented to both Houses of Parliament by Command of Her Majesty (Edinburgh: Her Majesty's Stationery Office, 1878), part 2, 3:318.
67. David Murray, *Lord Kelvin: As Professor in the Old College of Glasgow* (Glasgow: Maclehose, Jackson, 1924), pass.
68. W. T., Diary no. 3, 31 October 1846, pp. 41–43.
69. Ibid., 21, 28 and 29 November 1846, pp. 44 and 47.
70. J. D. Cormack, "Lord Kelvin, A Biographical Sketch," *Cassier's Magazine* 16 (May 1899): 151–52.
71. Ibid., p. 153.

Notes

Chapter 6

1. W. T. to Stokes, 5 November 1846, Stokes Collection, Add. MS 7656, CUL, (hereafter Add. MS 7656). Stokes to W. T., 10 November 1846; King, *Kelvin's Home,* pp. 236–37.

2. Forbes to W. T., 8 February 1847; W. T. to Stokes, 28 March 1847, Add. MS 7656. Fuller to W. T., 22 April 1847; James Thomson to W. T., 22 May 1847; Dr. Thomson to W. T., 25 May 1847.

3. W. T. to Dr. Thomson, 20 June 1847.

4. On 29 July 1847, Liouville wrote to W. T., "I have been told that he [Cayley] was giving up mathematics, that he wished to be an attorney. This would be a truly unhappy thing for science." W. T. to Dr. Thomson, 28 May 1847; W. T. to James Thomson, 11 June 1847, copy. See also Thompson, *Life,* 1:202.

5. W. T., 1845–1863 Diary, Diary no. 3, 31 October–28 November 1846, pp. 41–47. Diary no. 3 does not include January 1858 to 26 June 1859, which is Diary no. 4.

6. W. T., *Math. and Phys. Papers,* 1:76–80.

7. W. T. to Forbes, 29 December 1846, W. T.'s copy.

8. W. T., Diary no. 3, 29 March 1847, pp. 52–53; Stokes to W. T., 1 April 1847.

9. G. G. Stokes, *Mathematical and Physical Papers,* 5 vols. (Cambridge, 1880–1905), 1:157–87.

10. Stokes to W. T., 1 and 10 April 1847.

11. W. T. to Stokes, 20 December 1857, Add. MS 7656.

12. W. T., *Math. and Phys. Papers,* 1:107. In this volume are included references to the whole series "Notes on Hydrodynamics," appearing in *Cambridge and Dublin Mathematical Journal* (hereafter cited as *Camb. and Dub. Math. Jour.*).

13. W. T. to Stokes, 20 October 1847, Add. MS 7656.

14. W. T. to Stokes, 7 April 1847, Add. MS 7656.

15. W. T., *Elec. and Mag.,* pp. 468–70.

16. Ibid., pp. 468–69.

17. Ibid., p. 469.

18. Ibid., p. 423.

19. See Mary B. Hesse, *Models and Analogies in Science* (Notre Dame, Indiana: University of Notre Dame Press, 1970) for a very useful philosophical discussion of this question of which this chapter is a historical example. Dr. Hesse's book has many useful references for those interested in following the philosophical aspects of the concepts.

20. W. T., *Elec. and Mag.,* p. 470.

21. W. T. to Dr. Thomson, 28 May 1847; W. T., Diary no. 3, 22 June 1847, p. 66; W. T. to Dr. Thomson, 1 July 1847; "On Electrical Images," *Report of the British Association* (1847), Transactions of the Sections, pp. 6–7; "On the Electric Currents by which the Phenomena of Terrestrial Magnetism may be produced," pp. 38–39.

22. Fischer to W. T., 15 and 18 June 1847; W. T. to Dr. Thomson, 1 July 1847.

23. W. T. to Dr. Thomson, 1 July 1847.

24. Forbes to W. T., 1 July 1847; Thompson, *Life,* 1:199; Forbes to W. T., 15 July 1847; W. T. to Dr. Thomson, 15 August 1847.

25. W. T., Diary no. 3, 19 August 1847, pp. 79–80.

26. *Transactions of the Royal Society of Edinburgh* 16 (1849): 541–74.

27. J. P. Joule, "On the Calorific Effects of Magneto-Electricity, and on the Mechanical Value of Heat," in *Scientific Papers of James Prescott Joule,* 2 vols. (London: Physical Society of London, 1884–1887), 1:123.

28. Ibid., p. 122.

29. Joule to W. T., 29 June 1847. The papers were: "On the Calorific Effects of Magneto-Electricity, and on the Mechanical Value of Heat," ibid., pp. 123–59; and "On the Changes of Temperature Produced by the Rarefaction and Condensation of Air," pp. 171–89, in which Joule states his opposition to the popular view of heat as expressed by Clapeyron and Carnot, p. 188.

Notes

30. W. T. to Dr. Thomson, 1 July 1847; to James Thomson, 12 July 1847, copy; James Thomson to W. T., 24 July 1847.

31. W. T., *Math. and Phys. Papers,* 1:100–106. In a paper by S. G. Brush, "The Wave Theory of Heat: A Forgotten Stage in the Transition from the Caloric Theory to Thermodynamics," *British Journal for the History of Science* 5, no. 18 (1970): 165–66, he suggests that Thomson did not read the literature and thus was misled into believing the caloric theory. The evidence, I believe, indicates that Thomson changed his mind since he gradually was converted to the dynamical theory.

32. W. T., *Math. and Phys. Papers,* 1:100.

33. Ibid., p. 102.

34. Joule to W. T., 6 October 1848.

35. W. T. to Joule, 27 October 1848, badly faded copy.

36. W. T., *Math. and Phys. Papers,* 1:113–54, 100.

37. Ibid., pp. 116–17.

38. Ibid., pp. 114 and 118.

39. Ibid., p. 119.

40. W. T. to Dr. Thomson, 26 July 1847.

41. James Thomson to W. T., 29 July 1847; W. T. to Elizabeth King, 4 August 1847.

42. W. T. to Dr. Thomson, 15 and 22 August, 5 September 1847.

43. W. T. to Dr. Thomson, 22 August and 5 September 1847.

44. W. T., Diary no. 3, 21 September 1847, p. 80.

45. Ibid., 1 July 1846, p. 21; 22 July 1848, p. 95.

46. Ibid., pp. 75, 78, and 79.

47. W. T., *Math. and Phys. Papers,* 1:156–57, 164.

48. King, *Kelvin's Home,* pp. 241–42; Ellis to W. T., 17 January 1849.

49. James Thomson, *Collected Papers in Science and Engineering* (Cambridge: Cambridge University Press, 1912), p. xl.

50. *Camb. and Dub. Math. Jour.* 2 (1847): 61–64; W. T. to Stokes, 4 February 1849, Add. MS 7656.

51. W. T., *Elec. and Mag.,* pp. 344–45.

52. A broader historical view of the relation between the experimental and the physical was given by Thomas S. Kuhn in his 1972 George Sarton Memorial Lecture "Mathematical vs. Experimental Traditions in the Development of Physical Science." The essay was published in Thomas S. Kuhn, *Essential Tension, Selected Studies in Scientific Tradition and Change* (Chicago: University of Chicago Press, 1977). It helped me to see this relationship more clearly. See also H. I. Sharlin, *Convergent Century,* introduction.

Chapter 7

1. Niven, *Scientific Papers,* 2:401.

2. Sabine to W. T., 21 March 1849; W. T. to Dr. Thomson, 4 April 1844.

3. W. T., "A Mathematical Theory of Electricity," in *Elec. and Mag.,* pp. 409–30.

4. Ibid., p. 344.

5. Ibid., p. 365.

6. W. T., 1845–1863 Diary, Diary no. 3, pp. 136–37; W. T., *Elec. and Mag.,* p. 388; W. T. to Stokes, 25 January 1850, Add. MS 7656. Diary no. 3 does not include January 1858 to 26 June 1859, which is Diary no. 4.

7. W. T. to Stokes, 27 October 1849, Add. MS 7656.

8. Stokes to W. T., 29 October 1849.

9. W. T., *Elec. and Mag.,* p. 351.

10. Niven, *Scientific Papers,* 2:305.

11. W. T., Diary no. 3, 17 May 1850 [inserted between pp. 149 and 151].

12. Fischer to W. T., 4 November 1851.

Notes

13. W. T. to Stokes, n.d. [1849], Add. MS 7656. For a view of Kelvin's religious attitudes, see David B. Wilson, "Kelvin's Scientific Realism: the Theological Context," *Philosophical Journal* 11, no. 2 (July 1974): 41–60.

14. Hopkins to W. T., 9 February 1849.

15. Stokes to W. T., 12 February 1849.

16. W. T. to Stokes, Wednesday morning, n.d. [1849], Add. MS 7656; W. T. to Cookson, 28 November 1847.

17. W. T. to Stokes, Wednesday morning, n.d. [1849], Add. MS 7656; Stokes to W. T., 16 February 1849.

18. W. T. to Stokes, 20 February 1849, Add. MS 7656.

19. W. T. to Stokes, 21 March 1849, Add. MS 7656; Forbes to W. T., 21 March 1849; Blackburn to W. T., 5 June 1849; W. T., Diary no. 3, 6 September 1851, p. 159; 6 June 1851, p. 166.

20. S. P. Thompson, *Life,* 1:480.

21. Joule to W. T., 26 March 1850.

22. W. T., draft of "Dynamical Theory of Heat," [p.1].

23. Ibid., [pp. 2–3, 6–7].

24. Ibid., [pp. 10–11].

25. Ibid., [p. 12].

26. Ibid., [pp. 15–17].

27. Ibid., [p. 21]; W. T., *Math. and Phys. Papers,* 1:175.

28. Ibid., 1:176 and 178.

29. Ibid., 1:179.

30. Thompson, *Life,* 1:292; *Math. and Phys. Papers,* 1:181; "draft," [p. 3].

31. *Math. and Phys. Papers,* 1:175–76, 181–83.

32. Ibid., 1:188–89.

33. Ibid., 1:191.

34. Ibid., 1:183.

35. Joule to W. T., 28 April 1851; 1 and 6 May 1852.

36. Joule to W. T., 19 January 1854.

37. Joule to W. T., 13 August 1852.

38. W. T. to Joule, 25 April 1851, copy.

39. *Math. and Phys. Papers,* 1:333–455.

40. For a complete bibliography, see Thompson, *Life,* 2:1223–74.

41. W. T. to Stokes, 21 April 1851, 25 February 1851, Add. MS 7656.

42. In a letter to W. T. (21 August 1848), Stokes wrote that he had read two of Thomson's papers at the 1848 meeting of the British Association and that those present made no remarks. He doubted that many were able to follow the papers, which were "On the Equilibrium of Magnetic or Diamagnetic Bodies of any Form under the Influence of the Terrestrial Magnetic Force," *Report of the British Association,* Transactions of the Sections, pp. 8–9, and "On the Theory of Electromagnetic Induction," ibid., pp. 9–10.

43. See L. Campbell and W. Garnett, *The Life of James Clerk Maxwell* (London, 1882) for a discussion of Maxwell's close relationship to Blackburn's bride, Jemima Wedderburn, and her mother; see also p. 132.

44. Tyndall and Knoblauch, "Second Memoir on the Magneto-optic Properties of Crystals, and the relation of Magnetism and Diamagnetism to Molecular arrangement," *Philosophical Magazine* 37 (July 1850): 1–33; W. T. to Stokes, 5 and 15 July 1850, Add. MS 7656. *Report of the British Association* (1850), Transactions of the Sections, p. 23.

45. With the letter, Thomson included his paper "On the Forces experienced by small Spheres under Magnetic Influence; and on some of the Phenomena presented by Diamagnetic Substances," *Camb. and Dub. Math. Jour.* 2 (1847): 230–35. Diary no. 3, 14 August 1850, p. 154; see also Campbell, *Life of Maxwell,* pp. 144–45.

46. Maxwell to W. T., 20 February 1854; 13 November 1854; 15 May 1855, Add.

Notes

2766, CUL. See also J. Larmor, *Origins of Clerk Maxwell's Electric Ideas: as described in familiar letters to William Thomson* (Cambridge: Cambridge University Press, 1937).

47. Maxwell to W. T., 13 September 1855.
48. Niven, *Scientific Papers*, 1:156–58.
49. Ibid., p. 160.
50. Ibid., pp. 207–8.
51. Ibid., p. 452; Maxwell to W. T., 10 December 1861, ADD. 2766, CUL; Kelvin, *Baltimore Lectures* (London, 1904), Appendix F., pp. 569–77.
52. Niven, *Scientific Papers*, 2:306.
53. Ibid.
54. J. C. Maxwell, *A Treatise on Electricity and Magnetism*, reprinted from 3d. ed., 2 vols. (London: Oxford University Press, 1904), 1:viii–ix.
55. See Edmund T. Whittaker, "The Followers of Maxwell," in *A History of the Theories of Aether and Electricity*, vol. 1.
56. Ibid., 1:ix.
57. W. T., *Elec. and Mag.*, p. 1.

Chapter 8

1. *Times* (London), 30 July 1866, p. 8.
2. Quoted in Thompson, *Life*, 1:233.
3. W. T. to Stokes, 31 July 1852, Add. MS 7656.
4. *Memoir and Scientific Correspondence of the Late Sir George Gabriel Stokes, Bart.*, selected and arranged by Joseph Larmor, 2 vols. (Cambridge: Cambridge University Press, 1907), 1:14–15, 71.
5. Thompson, *Life*, 1:238 and 305.
6. J. D. Forbes to W. T., 20 May 1855.
7. W. T. to Stokes, 20 February 1854, Add. MS 7656. Stokes to W. T., 24 February 1854.
8. W. T. to Stokes, 5 February 1848.
9. Leo Koenigsberger, *Hermann von Helmholtz*, trans. F. A. Welby, with a preface by Lord Kelvin (New York: Dover, 1965), an unaltered and unabridged version of the 1906 Oxford edition, p. 145.
10. Thompson, *Life*, 1:501 and 504.
11. *Times* (London), 12 November 1856, p. 7.
12. H. I. Sharlin, *The Making of the Electrical Age: From the Telegraph to Automation* (London and New York: Abelard-Schuman, 1963), p. 34.
13. Charles F. Briggs and Augustus Maverick, *The Story of the Telegraph and A History of the Great Atlantic Cable* (New York: Rudd & Carleton, 1858), p. 222.
14. W. T., *Math. and Phys. Papers*, 2:138; Michael Faraday, *Experimental Researches in Electricity*, 3 vols. (London: Richard & John E. Taylor, 1839–1855), 3:512.
15. Stokes to W. T., 16 October 1854.
16. W. T., "On the Theory of the Electric Telegraph," in *Math. and Phys. Papers*, 2:61.
17. W. T., "On Peristaltic Induction of Electric Currents in Submarine Telegraph Wires," ibid., p. 78.
18. W. T., "Letters on 'Telegraph to America,'" ibid., p. 92; see Sir David Brewster, "The Atlantic Telegraph," *North British Review* 29 (November 1858): 546–53. Brewster heaped praise on Thomson and his galvanometers, but Thomson, in a letter to J. P. Nichol on 15 March 1859, criticized Brewster's unquestioned acceptance of Whitehouse's ideas.
19. W. T., "On Practical Methods for Rapid Signalling by the Electric Telegraph," ibid., p. 103.
20. See Derek J. Price, "Lord Kelvin, hero of the Atlantic telegraph," *Times Educational Supplement*, 1956, p. 1348.

21. *Math. and Phys. Papers,* 2:63–66.

22. Ibid., pp. 63 and 65.

23. W. T. to Stokes, 28 October 1854, Add. MS 7656.

24. *Math. and Phys. Papers,* 2: 67–68.

25. *Athenaeum,* 30 August 1856, p. 1093.

26. *Math. and Phys. Papers,* 2: 94–98.

27. Ibid., p. 100.

28. *Athenaeum,* 8 November 1856, p. 1371.

29. W. T., 1845–1863 Diary, Diary no. 3, 6 September 1850, p. 159; 13 September 1850, pp. 162–63.

30. *Math. and Phys. Papers,* 2:54 and 72.

31. Ibid., p. 41.

32. W. T., "On the Electro-Statical Capacity of a Leyden Phial and of a Telegraph Wire Insulated in the Axis of a Cylindrical Conducting Sheath," in *Elec. and Mag.,* 2d. ed. (London, 1884), p. 38 and p. 38n.

33. T. H. Huxley, "Professor Tyndall," *Nineteenth Century* 35 (January 1894): 5–6.

34. Thomson wrote a 36-page letter to Stokes on the subject, 29 April 1859. See Kelvin Papers, University of Glasgow. W. T. to Nichol, 31 May 1859, Kelvin Papers, University of Glasgow.

35. E. B. Bright and Charles Bright, *The Life Story of the late Sir Charles Tilston Bright Civil Engineer: with which is incorporated the Story of the Atlantic Cable, and the first Telegraph to India and the Colonies,* 2 vols. (Westminster: Archibald Constable; n.d. [1898]). 1:121, 124–25.

36. Ibid., p. 162.

37. J. W. Brett to W. T., 29 December 1856.

38. Bright, *Charles Tilston Bright,* 1:192–96.

39. *Times* (London), 26 August 1857, p. 10.

40. W. T., *Math. and Phys. Papers,* 2:154.

41. Ibid., pp. 103–6, 101.

42. Ibid., p. 105.

43. Ibid.

44. Ibid., p. 102.

45. Ibid., p. 125.

46. *Times* (London), 3 April 1858, p. 7.

47. Thompson, *Life,* 1:353–54.

48. Ibid., *Report of the Joint Committee Appointed by the Lords of the Committee of Privy Council for Trade and the Atlantic Telegraph Company to Inquire into the Construction of Submarine Telegraph Cables; Together with the Minutes of Evidence and Appendix.* Presented to Both Houses of Parliament by Command of Her Majesty (London, 1861), p. 113.

49. *Times* (London), 15 July 1858, p. 10.

50. Thompson, *Life,* 1:362.

51. Ibid., p. 365.

52. Lampson to W. T., 22 October 1858.

53. W. T. to Nichol, 18 March 1859.

54. *Report of the Joint Committee,* p. ix.

55. Ibid., pp. v, xii, and xiii.

56. *Times* (London), 30 July 1866, p. 8.

Chapter 9

1. P. G. Tait to W. T., 12 December 1861.

2. Thompson, *Life,* 1:412–13; W. T. to Stokes, 16 January 1861, Stokes Papers, CUL., Add. MS 7656.

Notes

3. Thompson, *Life*, 1:414.

4. W. T. to Jenkin, 19 April 1861, collection of J. K. Bottomley.

5. W. T. to Jenkin, 20 April 1861, ibid.

6. James Thomson to W. T., 26 April 1861.

7. C. G. Knott, *Life and Scientific Work of Peter Guthrie Tait* (Cambridge: Cambridge University Press, 1911), pp. 177–78.

8. Ibid., p. 16.

9. Thompson, *Life*, 1:450.

10. Tait to W. T., 12 December 1861.

11. Ibid.

12. Tait to W. T., 25 and 26 December 1861.

13. W. T. to Tait, 8 September 1862, collection of Margaret Tait.

14. Thompson, *Life*, 1:442.

15. Tait to W. T., 28 and 26 December 1861.

16. W. T., notebook on "Natural Philosophy," pp. 4–5.

17. "Hopkins Mss., 'Hydrostatics,' " pp. 33–35.

18. W. T., "Towards Friction & Cohesion," in Green book, p. 26.

19. Thompson, *Life*, 1:241–45.

20. P. G. Tait, *The Position and Prospect of Physical Science, a Public Inaugural Lecture, Delivered on November 7, 1860* (Edinburgh: Edmonston & Douglas, 1860), pp. 5, 7, and 34.

21. W. T., "Electrical Measurement," in *Popular Lectures and Addresses*, 1:455–59.

22. Tait to W. T., 16 January 1862.

23. J. Tyndall, *Fragments of Science*, 2 vols. (New York: D. Appleton, 1897), 1:380; J. Tyndall, "Remarks on an article entitled 'Energy' in 'Good Words,' " *Philosophical Magazine* 25 (March 1863): 221.

24. P. G. Tait, "Reply to Prof. Tyndall's *Remarks on a paper on 'Energy' in 'Good Words,' " Philosophical Magazine* 25 (April 1863): 264.

25. J. Tyndall, "Remarks on the Dynamical Theory of Heat in a letter to William Thomson," *Philosophical Magazine* 25 (May 1863): 369. W. Thomson, "Note on Professor Tyndall's 'Remark on the Dynamical Theory of Heat,' " *Philosophical Magazine* 25 (June 1863): 429.

26. W. T. to James Thomson, 28 January 1864, copy.

27. W. T., "Notebook & letters re: new ordinance at University of Glasgow." The notebook contained drafts of letters and statements as well as letters to Thomson.

28. Jenkin to W. T., 17 January 1862, collection of J. K. Bottomley; "Notebook & letters."

29. "Notebook & letters"; see also W. T., "The Bangor Laboratories," *Popular Addresses and Lectures*, 2:483–86.

30. W. T. to James Thomson, 17 March 1855, copy.

31. W. T. to Stokes, 7 February 1860, Add. MS 7656.

32. W. T. to Stokes, 26 November 1862, Add. MS 7656.

33. Tait to W. T., 15 January 1862.

34. Tait to W. T., 20 January 1862.

35. W. T., "Towards Friction & Cohesion," p. 1.

36. Thompson, *Life*, 1:434; Tait to W. T., 15 and 23 January, 4 March 1862.

37. Tait to W. T., 15 January 1862.

38. "Towards Friction & Cohesion," pp. 35 and 40.

39. Ibid., pp. 29–34.

40. W. T., "Beginnings of T & T," in Green book opposite p. 6.

41. Ibid., p. 1.

42. Thomson and Tait, *Treatise on Natural Philosophy* (Oxford, 1867), 1:178. For a version of Herbert Spencer's misunderstanding connected with the quotation from Thomson and Tait, see Knott, *Tait*, pp. 278–82.

43. "Beginnings of T & T," p. 5 and opposite.

44. Thompson, *Life,* 1:414–15. The Green books were small enough to fit into a jacket pocket and were brought out at any spare moment. They were less complete in later years, used more as scratch pads, and could be only understood by Thomson. By being less thorough in the later books, he was not keeping a usable record for later research. Thomson left approximately 268 Green books. In 1862, they were serving as records and as a convenient form for exchanging ideas.

45. W. T. to Tait, 14 February 1862.

46. A new edition was published in 1879 and was called part one. In 1883, part two was published. From 1886 to 1912, there were many editions, all substantially the same as the 1879 and 1883 editions. In 1962, Dover published parts one and two in one volume under the title *Principles of Mechanics and Dynamics.*

47. Tait to W. T., 13 May 1864.

48. Ibid.

49. Helmholtz to W. T., 14 December 1862.

50. T. & T., *Treatise,* p. vi.

51. Ibid., pp. 310 and 305.

52. Ibid., p. 311.

53. Ibid.

54. Ibid., p. 341.

55. Ibid., p. vi.

56. Ibid., pp. 337–39.

57. "Beginnings of T & T," opposite p. 1 [in Tait's handwriting]. T & T, *Treatise,* p. 178.

58. Ibid., pp. 179–80; "Beginnings of T & T," p. 1, back of flyleaf.

59. Knott, *Tait,* p. 202.

60. *Athenaeum,* 5 October 1867, p. 433.

61. Niven, *Scientific Papers,* 2:782.

62. *British Quarterly Review* 38 (July 1863): 256.

Chapter 10

1. W. T., *Math. and Phys. Papers,* 1:39–47.

2. Thompson, *Life,* 1:184–87.

3. W. T., "On a Universal Tendency in Nature to the Dissipation of Mechanical Energy," in *Math. and Phys. Papers,* 1:511–14. For a discussion of an alternative view by Tait, see P. M. Heiman, "The Unseen Universe: Physics and the Philosophy of Nature in Victorian Britain," *British Journal for the History of Science* 6, no. 21 (1972): 73–79.

4. W. T., *Math. and Phys. Papers,* 2:1–25.

5. In a note added in 1882, W. T. reversed himself and opted for the first theory, which stated that the sun was a heated body losing heat. See *Math. and Phys. Papers,* 2:3.

6. Ibid., p. 37.

7. *Cosmos* 6 (1855): 659.

8. Ibid.

9. W. T., *Math. and Phys. Papers,* 2:38.

10. Ibid., p. 40.

11. Ibid., 3: 295–311.

12. Ibid., p. 296.

13. Ibid.

14. W. T., "On the Rigidity of the Earth; Shiftings of the Earth's Instantaneous Axis of Rotation; and Irregularities of the Earth as a Timekeeper," Ibid., 3:312.

15. W. T., "On the Age of the Sun's Heat," in *Popular Addresses and Lectures,* 1:356–57.

16. W. T., "On the Mechanical Energies of the Solar System," in *Math. and Phys. Papers,* 2:24–25.

Notes

17. W. T., *Popular Addresses and Lectures*, 1:375.
18. William Hopkins, "Physical Theories of the Phenomena of Life," part 1, *Fraser's Magazine* 61 (1860): 739, 742–43.
19. Ibid., pp. 748–49.
20. T. H. Huxley, *Autobiography and Selected Essays*, ed. Ada L. F. Snell (Boston: Houghton Mifflin, 1909), p. 13.
21. William Irvine, "Thomas Henry Huxley," in *British Writers and Their Work: No. 2* (Lincoln, Nebr.: University of Nebraska Press, 1963), p. 96.
22. W. T., *Popular Addresses and Lectures*, 2:6–9.
23. T. H. Huxley, "The Anniversary of the President," *Quarterly Journal of the Geological Society of London* 25 (1869): lii–liii.
24. W. T., "Of Geological Dynamics," *Popular Addresses and Lectures*, 2:108–10.
25. W. T., "On Geological Time," *Popular Addresses and Lectures*, 2:44.
26. Huxley, "Address," p. xxxviii.
27. Ibid., pp. xlvi-xlvii.
28. Ibid., p. xlviii.
29. Ibid., p. lii; W. T., "Geological Dynamics," *Popular Addresses and Lectures*, 2:98.
30. W. T., "Geological Time," ibid., pp. 44–46.
31. ———, "Mathematics versus Geology," *Pall Mall Gazette* 9 (1869): 11–12.
32. A. R. Wallace, "The Measurement of Geological Time I," *Nature* 1 (1870): 399–401; James Marchant, *Alfred Russel Wallace: Letters and Reminiscences* (New York and London: Harper & Bros., 1916), pp. 198 and 220.
33. W. Boyd Dawkins, "Geological Theory in Britain," *Edinburgh Review*, American ed. 131 (1870): 22–23.
34. Ibid., p. 24.
35. W. T., "Geological Time," *Popular Addresses and Lectures*, 2:46.
36. Quoted in Joseph Larmor, ed., *Memoirs*, 1:168.
37. W. T., "Presidential Address to the British Association, Edinburgh, 1871," in *Popular Addresses and Lectures*, 2:203–4, 200.
38. Ibid., pp. 198 and 202.
39. Leo Koenigsberger, *Hermann von Helmholtz*, p. 145; Thompson, *Life*, 1:533.
40. Ibid., p. 577.

Chapter 11

1. W. T. to James Thomson, 21 September 1870, copy.
2. Thompson, *Life*, 2:586, 588–89.
3. Ibid., p. 594.
4. W. T. to Stokes, 20 July 1881, Add. MS 7656; Thompson, *Life*, 2:614–15.
5. W. T. to Stokes, 25 May 1872, Add. MS 7656.
6. W. T. (T. Tatlock) to Stokes, 6 October 1871, Add. MS 7656; James Thomson to George Darwin, 25 June 1880, Kelvin Papers, University of Glasgow.
7. W. T. (T. Tatlock) to Stokes, 6 October 1871, Add. MS 7656; W. T. to Tait, 28 February 1873, collection of Margaret Tait.
8. Thompson, *Life*, 2:596 and 598; W. T., *Popular Lectures and Addresses*, 2:132–205.
9. The University moved to a new site in November 1870.
10. Thompson, *Life*, 2:613.
11. Ibid., p. 597.
12. Ibid., p. 620.
13. Ibid., p. 616; Mrs. Hugh Blackburn, *Birds Drawn from Nature* (Edinburgh: Edmonston & Douglas, 1862); Thompson, *Life*, 2:616.
14. Thompson, *Life*, 2:637–38.
15. Ibid., p. 639.

Notes

16. Ibid., pp. 644–45; *Modern Society,* 11 November 1899:14.

17. Fanny Thomson was born on 12 September 1837. This information was included in a letter to H. I. Sharlin from P. Graham Blandy, dated 22 June 1968; Thompson, *Life,* 2:646.

18. Tait to W. T., 8 September 1887, collection of Margaret Tait.

19. W. T., "Navigation," in *Popular Lectures and Addresses,* 3:1–138. Thompson, *Life,* 2:657.

20. W. T., "Terrestrial Magnetism and The Mariner's Compass," in *Popular Lectures and Addresses,* 3:288–95.

21. Ibid., pp. 304–12; A memoir of W. T.'s Compass [1879?] in his secretary's handwriting with notes by W. T., pp. 1–2. Collection of J. K. Bottomley; Robert John Strutt, *John William Strutt, Third Baron Rayleigh,* an augmented edition with annotations by the author and foreword by John N. Howard (Madison and Milwaukee: University of Wisconsin Press, 1968), p. 78.

22. Memoir of W. T.'s Compass, pp. 2–3.

23. W. T., *Popular Lectures and Addresses,* 3:298–300.

24. *Nation* 23 (23 November 1876): 310. The foreign judges were paid $1,000, and the American judges, $600.

25. Thomas E. Slattery to H. I. Sharlin, 11 December 1967. Slattery is archival examiner of the Department of Records, Philadelphia, Pa.; *Report of the British Association* (1876), Transactions of the Sections, p. 2.

26. F. A. Walker, ed., *United States Centennial Commission International Exhibition, 1876. Reports and Awards, Groups XXI–XXVII* (Washington, 1880), Group 25: 7:131.

27. W. T. to Tait, 11 July 1876, collection of Margaret Tait; *Report of the British Association* (1876), Transactions of the Sections, p. 1.

28. Ibid.; W. T. to Tait, 11 July 1876.

29. *Report of the British Association* (1876), Transactions of the Sections, pp. 1–3.

30. *New York Times,* 2 October 1876, p. 4; *North American Review* 122 (January 1876): 88–123.

31. Leo Koenigsberger, *Hermann von Helmholtz,* p. 349.

32. A favorite term of Thomson's.

33. W. T. to Jenkin, 16 November 1861; Jenkin to W. T., 17 November 1861, collection of J. K. Bottomley; the 1861 Electrical Standards Committee consisted of Williamson, Wheatstone, Thomson, Miller (of Cambridge), Matthiessen, and Jenkin. In 1862, Esselbach, C. Bright, Maxwell, C. W. Siemens, and Balfour Stewart were added.

34. Fleeming Jenkin, ed., *Reports of the Committee on Electrical Standards Appointed by the British Association for the Advancement of Science* (London: F. N. Spon, 1873), pp. 1 and 41.

35. For a description of the use of measurement in modern physical science, see T. S. Kuhn, "The Function of Measurement in Modern Physical Science," in H. Woolf, ed., *Quantification: A History of the Meaning of Measurement in the Natural and Social Sciences* (Indianapolis: Bobbs-Merrill, 1961), pp. 31–63.

36. W. T., *Popular Addresses and Lectures,* 1:87 and 80.

37. W. T., Notebook on "Natural Philosophy," p. 1, 5 November 1840. This set of notes begins at the back of the notebook.

38. W. T., "On an Absolute Thermometric Scale founded on Carnot's Theory of the Motive Power of Heat, and calculated from Regnault's Observations," *Math. and Phys. Papers,* 1:102.

39. W. T., "Notes at a Meeting," in Green book [1860's], pp. 3–7.

40. W. T. to Stokes, 1 January 1873, Stokes Papers, CUL.

41. W. T., "Applications of the Principle of Mechanical Effect to the Measurement of Electro-Motive Forces and of Galvanic Resistances, in Absolute Units," *Math. and Phys. Papers,* 1:490.

42. Jenkin, *Electrical Standards,* pp. 41–42.

43. W. T., "Electrical Measurement," *Popular Addresses and Lectures,* 1:440–43.

Notes

44. W. T., "Electrical Units of Measurement," ibid., pp. 97 and 87.
45. W. T., "Presidential Address to the Society of Telegraph Engineers, 1874," ibid., 2:218, 227, 222–24.
46. W. T., "Atmospheric Electricity," in *Reprint of Papers on Electrostatics and Magnetism,* 2d. ed. (London, 1884), p. 202; W. T., "Notes on Atmospheric Electricity," *Report of the British Association* (1860), Transactions of the Sections, p. 53; W. T. to Stokes, 20 July 1881, Add. MS 7656; W. T. to secretary of Royal Society, 17 March 1875; J. D. Everett to W. T., 5 February 1861.
47. *Report of the British Association* (1859), Transactions of the Sections, pp. 27–28.
48. W. T. to James Thomson, 2 July 1856, copy extract; W. T., Diary no. 3, 11 June 1856, p. 207. See also *Report of the British Association* (1856), Transactions of the Sections, pp. 17–18.
49. W. T., *Elec. and Mag.,* pp. 208–26; Dellmann to W. T., 28 February 1862.
50. Copied by Margaret Tait from P. G. Tait's private notebook, n.d. [1889?]. Collection of Margaret Tait.
51. Strutt, *Rayleigh,* p. 240; Tait to W. T., 14 April 1870.
52. W. T., *Popular Addresses and Lectures,* 3:209–23, 154–55.
53. Ibid., pp. 209–16; W. T. to James Thomson, 18 July 1868.
54. W. T. to General R. Strachey, 10 December 1875. A copy printed in "Tidal Observations in the Indian Seas," nos. 81–5m. Fort William, 14 March 1876, Resolution by the Government of India, Public Works Dept. Collection of J. K. Bottomley.
55. W. T., "The Tide Gauge, Tidal Harmonic Analyzer, and Tide Predict[o]r," *Minutes and Proceedings of the Institution of Civil Engineers* 65 (1881): 10; James Thomson, *Collected Papers,* pp. 452–54.
56. W. T. to Stokes, 9 July 1879, draft. Collection of J. K. Bottomley. See also *Min. Proc. Inst. Civil Eng.* 65: 30–31.
57. Roberts to W. T., 28 June 1879, collection of J. K. Bottomley; *Min. Proc. Civ. Eng.* 65, p. 33.
58. *Min. Proc. Civ. Eng.* 65, p. 60. The Roberts-Thomson correspondence is in the J. K. Bottomley collection; Edward Roberts, "Preliminary Note on a new Tide-Predict[o]r," *Proceedings of the Royal Society* 29 (19 June 1879): 198–201.
59. W. T. to Stokes, 9 July 1879, in Lady Thomson's handwriting. Collection of J. K. Bottomley.
60. *Min. Proc. Civ. Eng.* 65: 2–72.
61. W. T. to Stokes, 3 December 1863, Add. MS 7656; "Notes prepared for Messrs. Constable by Professor Fleeming Jenkin, October 1884," pp. 1–2, collection of J. K. Bottomley.
62. "Sir Wm. Thomson & others and the Atlantic Telegraph Company's copy opinion of H. H. Dodgson," 4 January 1868, MS; "Sir Wm. Thomson & others and the Atlantic Telegraph Company and Anglo American Telegraph Company, Agreement as to use of Patents," 1869, draft.
63. W. T. to Varley, 24 November 1880, draft. Collection of J. K. Bottomley; "Memo of Financial Arrangement agreed to by—Thomson & Jenkin with the Western & Brazilian Tel. Co.," 22 November 1880. Collection of J. K. Bottomley. See also Jenkin to W. T., 10 July 1880, collection of J. K. Bottomley.
64. W. T. to ?, 7 June 1883, copy.
65. Lady Thomson to George Darwin, 17 June 1881, Kelvin Papers, University of Glasgow.
66. Lady Thomson to George Darwin, 27 September 1879, Kelvin Papers, University of Glasgow; Margaret E. Gladstone, granddaughter of Elizabeth Thomson King and later Mrs. Ramsay MacDonald, quoted in Thompson, *Life,* 2:927.
67. J. J. Thomson, *Recollections and Reflections* (New York: Macmillan, 1937), pp. 99–100; Maxwell to Blore, 15 February 1871, Add. 7655, CUL.
68. W. T. to Stokes, 15 October 1879, Add. MS 7656. There are two such letters, the one not cited here is marked "(No 2)."

Notes

69. Maxwell to W. T., 13 November 1854, Add. 2766, CUL. Also see J. Larmor, *Maxwell's Electric Ideas.*
70. W. T. to Routh, 22 March 1882, copy.
71. W. T. to Stokes, 1 October 1879, Add. MS 7656; *Nation* (19 October 1882): 337.

Chapter 12

1. Daniel C. Gilman, *The Launching of a University and other papers, A Sheaf of Remembrances* (New York: Dodd, Mead, 1906), p. 75.
2. Simon Newcomb, "Abstract Science in America, 1776–1876," *North American Review* 122 (January 1876): 88–123.
3. Gilman to W. T., 8 January 1884; abstract of Gibbs to Gilman enclosed in above; W. T. to Gilman, 2 October 1883, copy.
4. Rowland to W. T., 8 January 1884.
5. Cyrus Field to W. T., 20 May 1884.
6. *New York Times,* 27 August 1884, p. 1; *Science* 4 (1884): 160; *Popular Science Monthly* 25 (1884): 700.
7. *American Journal of Science,* 3d. ser., 28 (1884): 300; *Science* 4 (1884): 271.
8. *New York Times,* 28 August 1884, p. 1, and 27 August 1884, p.1.
9. Ibid., 27 August 1884, p.1.
10. Thomson was standing in for Cayley, the vice-president.
11. *Popular Science Monthly* 25 (1884): 841.
12. W. T., *Popular Lectures and Addresses,* 1:225–59.
13. *Science* 4 (1884): 249.
14. *American Journal of Science* 28:305–6; *Science* 4:161.
15. *American Journal of Science* 28:306; *Science* 4:296.
16. Lady Thomson to George Darwin, 7 October 1884, Kelvin Papers, University of Glasgow.
17. *Times* (London), 30 September 1884, p. 5.
18. Diary of Henry Crew, 30 September 1884, Henry Crew Papers, Niels Bohr Library, American Institute of Physics.
19. Lady Thomson to George Darwin, 7 October 1884, Kelvin Papers, University of Glasgow.
20. *Science* 4 (1884): 349.
21. For a discussion of Michelson, Morley, and Thomson, see Lloyd S. Swenson, Jr., *The Ethereal Aether: A History of the Michelson-Morley-Mill Aether-Drift Experiments, 1880–1930,* with a foreword by Gerald Holton (Austin and London: University of Texas Press, 1972), pp. 76–81; Nathan Reingold, "Cleveland Abbe," *DSB.*
22. Hathaway's notes were reproduced by a device called the "papyrograph process." The edition of 300 copies was sold quickly. John C. French, *A History of the University founded by Johns Hopkins* (Baltimore: Johns Hopkins Press, 1946), p. 222.
23. Lord Kelvin, *Baltimore Lectures on Molecular Dynamics and The Wave Theory of Light: Founded on Mr. A. S. Hathaway's Stenographic Report of Twenty Lectures delivered in Johns Hopkins University, Baltimore, in October 1884: Followed by Twelve Appendices on Allied Subjects* (London, 1904), p. vii.
24. Ibid., pp. 6 and 9.
25. Ibid., p. 406.
26. For additional information, see Mary B. Hesse, *Models and Analogies in Science* (Notre Dame, Ind.: University of Notre Dame Press, 1966); H. I. Sharlin, *Convergent Century.*
27. Kelvin, *Baltimore,* pp. 34, 105, and 186.
28. Pierre Duhem, *The Aim and Structure of Physical Theory,* trans. Philip B. Wiener (Princeton: Princeton University Press, 1954), pp. 81 and 98.

29. Ibid., p. 81.

30. Kelvin, *Baltimore*, pp. 14, 12, 44, and 264–65.

31. Ibid., p. 45. A reading of the *Baltimore Lectures* reveals Thomson's fundamental complaint about Maxwell's theory. He did not reject the theory, but unlike most scientists, Thomson thought the theory was a blind alley. His objection was not due to an inability to make a model of the theory. See also W. T., *Popular Lectures and Addresses*, 2:547–48.

32. Kelvin, *Baltimore*, pp. 376–77.

33. *Nature* 31 (1885): 461–63, 601–3.

34. Ibid., p. 503. "Velocity of Electricity" was reprinted in *Math. and Phys. Papers*, 2:131–37.

35. Fitzgerald to W. T., 25 April 1885; *Nature* 32 (1885): 4–5.

36. W. T., *Popular Lectures and Addresses*, 2:547.

37. Diary of Henry Crew, 17 October 1884; *Science* 4 (1884): 437–38.

38. Ibid., p. 438; Kelvin, *Baltimore*, p. 22.

39. *Nature* 31 (1885): 461.

40. R. J. Strutt, *John William Strutt*, p. 145.

41. Kelvin, *Baltimore*, p. v.

42. W. T., "Ether, Electricity, and Ponderable Matter," *Math. and Phys. Papers*, 3:501–2.

43. W. T., "Steps Towards a Kinetic Theory of Matter," *Popular Lectures and Addresses*, 1:240.

44. Strutt, *John William Strutt*, p. 147. W. T. to master of Peterhouse, 25 November 1884.

45. W. T. to Stokes, 30 November 1884, Add. MS 7656; Lady Thomson to George Darwin, 12 December 1884, Kelvin Papers, University of Glasgow.

46. W. T. to George Darwin, 26 December 1884, Kelvin Papers, University of Glasgow.

47. Ibid., 30 December 1884, Kelvin Papers, University of Glasgow.

48. W. T., "On Vortex Atoms," *Math. and Phys. Papers*, 4:1.

49. Ibid., p. 3.

50. Ibid., p. 2. Gravity: a question which also troubled Einstein. His answer was the general theory of relativity.

51. W. T., "Elasticity Viewed as Possibly a Mode of Motion," *Popular Lectures and Addresses*, 1:153.

52. Tait to W. T., 20 December 1883; W. T., *Math. and Phys. Papers*, 3:509.

53. Ibid., pp. 510–11.

54. Robert H. Silliman, "William Thomson: Smoke Rings and Nineteenth-Century Atomism," *Isis* 54 (December 1963): 472.

Chapter 13

1. Thompson, *Life*, 2:920; Gunning to W. T., 24 March 1890.

2. Muir to Kelvin, 11 June 1892; list of members and constitution and rules of West of Scotland Liberal Unionist Association, 1892.

3. *Scotsman*, 8 March 1893.

4. As quoted in Thompson, *Life*, 2:1130.

5. Ibid., p. 906; George Fitzgerald, "Biographical Sketch of Lord Kelvin," in *Lord Kelvin, Professor of Natural Philosophy in the University of Glasgow 1846–1899* (Glasgow: James Maclehose, 1899), pp. 4–5.

6. Kelvin Papers, CUL.

7. Dr. Thomson to W. T., 16 May 1846.

8. S. P. Thompson to W. T., 20 April and 6 May 1887, collection of J. K. Bottomley.

9. For additional information on this scheme, see H. I. Sharlin, *Electrical Age*, chap. 7.

Notes

10. A copy of Kelvin's cable to Adams, 1 May 1893, is included in Adams correspondence to Kelvin, 9 June 1893; collection of J. K. Bottomley. Thompson, *Life*, 2:997.

11. Perry to Kelvin, 23 October 1894; Perry to Tait, 26 November 1894.

12. See also Joe D. Burchfield, *Lord Kelvin and the Age of the Earth* (New York: Science History Publications, 1975), pp. 135–37; Kelvin to Perry, 13 December 1894. Most of the correspondence between Perry and Tait and Perry and Kelvin—all tied together in Kelvin Papers, CUL—was printed in *Nature* 51 (1895): 224–27.

13. W. T., *Math. and Phys. Papers*, 6:54–56.

14. Stokes to Kelvin, 17 February 1896.

15. Kelvin to Stokes, 24 February 1896, Add. MS 7656; Kelvin to Stokes, telegram, 26 February 1896.

16. *Math. and Phys. Papers*, 6:55–56.

17. Ibid., p. 57.

18. Stokes to Kelvin, 28 February 1896.

19. Kelvin to Stokes, 9 March 1896, Add. MS 7656.

20. Stokes to Kelvin, 9 March 1896; Kelvin to Stokes, 12 March 1896, Add. MS 7656; Stokes to Kelvin, 12 March 1896.

21. Stokes to Kelvin, 13 March 1896.

22. Stokes to Kelvin, 16 [19], 17 March 1896; Kelvin to Stokes, 17 March 1896, Add. MS 7656.

23. Stokes to Kelvin, 26 March 1896. There are two letters with this date. The one that is quoted is typed. See also *Proceedings of the Royal Society* 59 (1895–96):332.

24. Stokes to Kelvin, 26 and 31 March 1896; Kelvin to Stokes, 5 April 1896, Add. MS 7656.

25. Stokes to Kelvin, 6 April 1896. Either this letter or Kelvin's answer is dated incorrectly. Kelvin to Stokes, 5 April 1896, Add. MS 7656.

26. Kelvin to Stokes, 24 April and 19 June 1896, Add. MS 7656.

27. Merz, *European Thought*, 2:192.

28. G. Fitzgerald to Kelvin, 13 March 1896.

29. Niven, *Scientific Papers*, 2:361.

30. Kelvin, "On the Motion Produced in an Infinite Elastic Solid by the Motion through the Space Occupied by it of a Body Acting on it Only by Attraction or Repulsion," *Baltimore*, p. 468.

31. Ibid., p. 485; Russell McCormmach, "H. A. Lorentz and the Electromagnetic View of Nature," *Isis* 61 (Winter 1970):490.

32. Kelvin, *Baltimore*, pp. 487–88, 492.

33. Kelvin, "On the Clustering Matter in Any Part of the Universe," ibid., p. 533.

34. Ibid., p. vii.

35. Ibid., p. 467.

36. Thompson, *Life*, 2:984.

37. *Nature* 68 (October 1903): 624.

38. W. T., *Popular Lectures and Addresses*, 2:491–92.

39. George Fitzgerald to Kelvin, 4 April 1896.

40. J. J. Thomson, *Recollections and Reflections*, pp. 50 and 51.

41. Thompson, *Life*, 2:1065.

42. Fitzgerald to Kelvin, 17 April 1896; *Baltimore*, p. 9.

43. Fitzgerald to Kelvin, 17 April 1896.

44. Fitzgerald to Kelvin, 14 February 1896. Kelvin's remarks are written on Fitzgerald's letter; Fitzgerald to Kelvin, 17 February 1896.

45. Fitzgerald to Kelvin, 12 February and 11 May 1896.

46. Fitzgerald to Kelvin, 25 November 1898.

47. Kelvin to Fitzgerald, copy, 28 November 1898, Kelvin Papers, University of Glasgow.

48. Fitzgerald to Kelvin, 29 November 1898.

Notes

49. Heaviside to Kelvin, 5 April 1904; Kelvin to Heaviside, 20 April 1904, Institution of Electrical Engineers.

50. Heaviside to Kelvin, 24 April 1904.

51. Stokes to Kelvin, 22 November, 30 and 31 December 1898; Kelvin to Stokes, 2 January 1899, Add. MS 7656.

52. Stokes to Kelvin, 7 January 1899.

53. Stokes to Kelvin, 18 February 1899, 9 January 1901.

54. Kelvin to Story, 24 June 1899, Kelvin Papers, University of Glasgow. In 1904, Kelvin was appointed chancellor of the University of Glasgow.

55. King, *Kelvin's Home,* p. 236; Kelvin to Heaviside, 14 October 1899, Institution of Electrical Engineers; Lady Kelvin to George Darwin, 6 February 1900, Kelvin Papers, University of Glasgow.

56. Kelvin to Tait, 23 June 1901, collection of Margaret Tait; Kelvin, "James Watt," in *Math. and Phys. Papers,* 6:345–62; Niven, *Scientific Papers,* 2:782.

57. *Math. and Phys. Papers,* 6:369.

58. *Philosophical Magazine* 1, 6th ser. (1901): 360; *Nature* 63 (1901): 445–47.

59. *New York Times,* 9 May 1902, p. 2.

60. Ibid., 20 April 1902, p.2; Thompson, *Life,* 2:1197; *In Memoriam: The Right Honorable William Thomson, Lord Kelvin,* pamphlet (New York: American Institute of Electrical Engineers, 1908), p. 21; *Nature* 65 (1902): p. 591.

61. Ibid.; *Harper's Weekly* 46 (1902): 538 and 543.

62. *New York Times,* 9 May 1902, p. 2; *Nature* 65 (1902): 531.

63. *New York Times,* 25 April 1902, p. 1; Stokes to W. T., 24 May 1882.

64. Thompson, *Life,* p. 1167.

65. Kelvin to Chrystal, 16 January 1905, Kelvin Papers, University of Glasgow.

66. *Math. and Phys. Papers,* 6:344.

67. Kelvin to Becquerel, 4 December 1903.

68. Kelvin to Green, 31 August 1906, Kelvin Papers, University of Glasgow; Becquerel to Kelvin, 4 October 1906. Henri Becquerel and his wife visited Lord and Lady Kelvin at Netherhall a short time after the debate more or less concluded. Becquerel and Kelvin talked of recent literary works that had been inspired by radioactivity, and Becquerel recalled a French verse from the seventeenth century; it was from act 3, scene 7 of Racine's *Athalie,* where the idea of transmutation of elements was taken figuratively.

69. *Times* (London), 9 August 1906, p. 3.

70. Ibid., 15 August 1906, p.4. A similar charge against Kelvin has been made recently; see S. G. Brush, "The Wave Theory of Heat: A Forgotten Stage in the Transition from the Caloric Theory to Thermodynamics," *British Journal for the History of Science* 5 (1970): 166.

71. *Times* (London), 20 August 1906, p. 6.

72. Kelvin to Armstrong, 13 September 1906.

73. J. J. Thomson, *Recollection,* p. 240; R. J. Strutt, *John William Strutt,* p. 239; Thomas S. Kuhn, *The Structure of Scientific Revolutions* (Chicago: University of Chicago Press, 1962), pp. 18–19.

74. A. S. Eve, *Rutherford, Being the Life and Letters of the Rt. Hon. Lord Rutherford, O.M.* (New York and Cambridge: Macmillan and Cambridge University Press, 1939), p. 107.

75. Kelvin, "On the Motion of Ether produced by Collision of Atoms on Molecules containing or not containing Electrons," *Philosophical Magazine,* 6th ser., 14 (1907):317–24.

76. *Math. and Phys. Papers,* 6:235–43; *Times* (London), 2 August 1907, p. 9.

77. Ronald W. Clark, *Einstein, The Life and Times* (New York and Cleveland: World, 1971), pp. 332–40.

78. Ibid., pp. 113 and 612.

79. *James Clerk Maxwell, A Commemoration Volume 1831–1931* (Cambridge: Cambridge University Press, 1931), p. 71.

Notes

80. *Times* (London), 2 August 1907, p. 9.

81. See Maxwell's numerous references to Kelvin's papers, e.g., Niven, *Scientific Papers*, 1:209.

82. T. S. Kuhn's *Structure of Scientific Revolutions* is a good example of how S. P. Thompson's biography of Kelvin established the Kelvin myth. Kuhn, in a discussion of X rays on p. 59, writes: "Lord Kelvin at first pronounced them an elaborate hoax." He cites Thompson, 2:1125. On p. 150 of *Structure,* Kuhn reiterates Thompson, 2:879 and 1023, without citing him, that Kelvin never accepted Maxwell's theory. Another example is Ernest Nagel, *The Structure of Science* (New York, 1961). On p. 114, Nagel writes that Kelvin "never felt entirely at ease with Maxwell's electromagnetic theory of light because he was unable to design a satisfactory mechanical model for it." No reference is cited, but Nagel could have cited Thompson, 2:1025; *Nature* 77 (1907): 177.

83. Thompson, *Life,* 2:879, 1023–24.

84. Ibid., pp. 1027–28.

85. Ibid., p. 1086.

INDEX

Index

Index

Index

Index

Index

Index

Index

Index